1983

A PRIMER OF
REAL FUNCTIONS

By

RALPH P. BOAS, JR.

THE

CARUS MATHEMATICAL MONOGRAPHS

Published by

THE MATHEMATICAL ASSOCIATION OF AMERICA

T HE CARUS MATHEMATICAL MONOGRAPHS are an expression of the desire of Mrs. Mary Hegeler Carus, and of her son, Dr. Edward H. Carus, to contribute to the dissemination of mathematical knowledge by making accessible at nominal cost a series of expository presentations of the best thoughts and keenest researches in pure and applied mathematics. The publication of the first four of these monographs was made possible by a notable gift to the Mathematical Association of America by Mrs. Carus as sole trustee of the Edward C. Hegeler Trust Fund. The sales from these have resulted in the Carus Monograph Fund, and the Mathematical Association has used this as a revolving book fund to publish the succeeding monographs.

The expositions of mathematical subjects which the monographs contain are set forth in a manner comprehensible not only to teachers and students specializing in mathematics, but also to scientific workers in other fields, and especially to the wide circle of thoughtful people who, having a moderate acquaintance with elementary mathematics, wish to extend their knowledge without prolonged and critical study of the mathematical journals and treatises. The scope of this series includes also historical and biographical monographs.

The following monographs have been published:

No. 1. Calculus of Variations, by G. A. BLISS

No. 2. Analytic Functions of a Complex Variable, by D. R. CURTISS

No. 3. Mathematical Statistics, by H. L. RIETZ

No. 4. Projective Geometry, by J. W. YOUNG

No. 5. A History of Mathematics in America before 1900, by D. E. SMITH and JEKUTHIEL GINSBURG (out of print)

No. 6. Fourier Series and Orthogonal Polynomials, by DUNHAM JACKSON

No. 7. Vectors and Matrices, by C. C. MacDUFFEE

No. 8. Rings and Ideals, by N. H. McCOY

No. 9. The Theory of Algebraic Numbers, Second edition, by HARRY POLLARD and HAROLD G. DIAMOND

The Carus Mathematical Monographs

NUMBER THIRTEEN

A PRIMER OF
REAL FUNCTIONS

By

RALPH P. BOAS, JR.

Professor Emeritus of Mathematics
Northwestern University

Published and Distributed by
THE MATHEMATICAL ASSOCIATION OF AMERICA

Complete Set ISBN 0-88385-000-1
Vol. 13 ISBN 0-88385-022-2

Printed in the United States of America

Current printing (last digit):

10 9 8 7 6 5 4 3 2 1

PREFACE

I. To the beginner. In this little book I have presented some of the concepts and methods of "real variables" and used them to obtain some interesting results. I have not sought great generality or great completeness. My idea is to go reasonably far in a few directions with a minimum amount of special terminology. I hope that in this way I have been able to preserve some of the sense of wonder that was associated with the subject in its early days but has now largely been lost. I hope also that someone who has read this book will be able to go on to one of the many more forbidding systematic treatises, of which there is no lack.

No previous knowledge of the subject is assumed, but the reader should have had at least a course in calculus. In general, each topic is developed slowly but rises to a moderately high peak; a reader who finds the slope too steep may skip to the beginning of the next section.

Since this is not a handbook, but more in the nature of a course of informal lectures, I have not been at all consistent either about the proportion of detailed proof to general discussion or about strict logical arrangement of material.

All phrases like "it is clear," "plainly," "it is trivial" are intended as abbreviations for a statement something like "It should seem reasonable, you should be able to supply the proof, and you are invited to do so." On the other hand, "It can be shown . . . " is usually to suggest that the proof is too complicated to give here, or depends on notions that are not discussed here, and that you are not expected to try to supply the proof yourself.

In stating definitions, I have frequently used "if" where

I should really have used "if and only if." For example, "If a set is both bounded above and bounded below, it is called *bounded*." This definition is to be understood to carry an additional clause, "and if it is not both bounded above and bounded below, it is not called bounded."

There are a number of exercises, some of which merely supply illustrative material, and some of which are essential parts of the book. An exercise that merely states a proposition is to be interpreted as a demand for a proof of the proposition. Answers to all exercises are given at the end of the book.

Paragraphs in small type deal either with peripheral material or with more difficult questions.

I apologize in advance for whatever mistakes the alert reader may be able to detect. None were intentionally included; nevertheless, the detection and rectification of mistakes is a good exercise, and fosters a healthy skepticism about the printed word.

II. To the expert.   Experts are not supposed to read this book at all; since this statement will doubtless be taken as an invitation for them to do so, I must explain what I have tried (and not tried) to do. I have set out to tell readers with no previous experience of the subject some of the results that I find particularly interesting. I have therefore tried to present the material that seemed essential for the results I had in mind, together with as much related material as seemed interesting and not too complicated. Since this is not a systematic treatise, I have deliberately tried not to introduce any concepts or notations, however significant or convenient, that I did not really need to use. I have omitted integration, reluctantly, because of the many technical details that are needed before one gets to the interesting results.

Since this is not a treatise it has not been written like one. The style is deliberately wordy. The axiom of choice is frequently used but never mentioned; this book is not the place to discuss philosophical questions, and, in any case, after Gödel's results, the assumption of the axiom of choice can do no mathematical harm that has not already been done. On the other hand, according to the more recent work of P. J. Cohen, by assuming the axiom of choice rather than its negative we are selecting one kind of mathematics rather than another, say Zermelonian rather than non-Zermelonian. With this selection, there is no point in avoiding the axiom of choice whenever it seems natural to use it, even in cases where it is known to be avoidable.

III. Acknowledgments.   I am indebted to my teachers, J. L. Walsh and D. V. Widder, for introducing me to this kind of mathematics; to M. L. Boas and to E. F. and R. C. Buck for criticizing early drafts of the book; and to H. M. Clark and H. M. Gehman for help with the proofreading. I am grateful to several people who have pointed out oversights or suggested improvements, and especially to Richard L. Baker, E. M. Beesley, H. P. Boas, A. M. Bruckner, G. T. Cargo, S. Haber, A. P. Morse, C. C. Oehring, J. M. H. Olmstead, J. C. Oxtoby, A. C. Segal, A. Shuchat, and H. A. Thurston.

* * * * * *

In preparing this revision, I have tried to resist the temptation to insert additional material; but a considerable number of references have been added to the notes.

RALPH P. BOAS, JR.

Northwestern University
March 1960
June 1965
March 1972
October 1979

TO MY EPSILONS

CONTENTS

CHAPTER 1

SETS

1. Sets. In order to read anything about our subject, the reader will have to learn the language that is used in it. I shall try to keep the number of technical terms as small as possible, but there is a certain minimum vocabulary that is essential. Much of it consists of ordinary words used in special senses; this practice has both advantages and disadvantages, but has in any case to be endured since it is now too late to change the language completely. Much of the standard language is taken from the theory of sets, a subject with which we are not concerned for its own sake. The theory of sets is, indeed, an independent branch of mathematics. It has its own basic undefined concepts, subject to various axioms; one of these undefined concepts is the notion of "set" itself.

From an intuitive point of view, however, we may think of a *set* as being a collection of objects of some kind, called its *elements*, or *members*, or *points*. We say that a set contains its elements, or that the elements belong to the set or simply are in the set. The normal usage of set, as in "a set of dishes" or "a set of the works of Bourbaki," is fairly close to what we should have in mind, although the second phrase suggests some sort of arrangement of the elements which is irrelevant to the mathematical concept. Sets may, for example, be formed of ordinary geometrical points, or of functions, or indeed of other sets. We shall use the words *class*, *aggregate* and *collection* interchangeably with set, especially to make complicated situations clearer: thus we may speak of a collection of aggregates of sets rather than a set of sets of sets.

If E is a set, a set H is called a *subset* of E if every element of H is also an element of E.* For example, if E is the set whose elements are the numbers 1, 2, 3, there are eight subsets of E. Three of them contain one element each; three contain two elements each; one is the set E itself (a subset does not have to be, in any sense, "smaller" than the original set); the eighth subset of E is, by convention, the *empty set*, which is the set that has no elements at all. If H is a subset of E we write $H \subset E$ or $E \supset H$; sometimes we say that E contains H or that E covers H. If H is a subset of E but is not all of E, we call H a *proper* subset of E.

We write $x \in E$ to mean that x is an element of E. We often say that x is in E, or that x belongs to E, or that E contains x, meaning the same thing. Since the elements of sets are usually things of a different kind from the sets themselves, we should distinguish between the element x and the set whose only element is x. It is often convenient to denote the latter set by (x). The notations $x \in E$ and $(x) \subset E$ mean the same thing.

A *space* is a set that is being thought of as a universe from which sets can be extracted. If Ω is a space and $E \subset \Omega$, the *complement* of E (with respect to Ω) is the set consisting of all the elements of Ω that are not elements of E. The complement of E is denoted by $C(E)$. For example, if Ω consists of the letters of the alphabet and E of the consonants (including y as a consonant), $C(E)$

---

*It is traditional to call sets "E," presumably because the French for "set" is "ensemble."

consists of the vowels. If, however,[†] E consists of the single letter a, $C(E)$ consists of the letters b, c, ..., z. If E consists of the entire alphabet, $C(E)$ is empty. If E is empty, $C(E) = \Omega$.

Exercise 1.1. Show that $C(C(E)) = E$.

Occasionally it is necessary to consider the complements of a set with respect to different spaces; in such cases, special notations will be used.

If E and F are two sets, there are two other sets that can be formed by using them, and that occur so frequently that they have special names. One of these sets is the *union* of the two sets, written $E \cup F$ (sometimes called their sum, and written $E + F$); it consists of all elements that are in E or in F (or in both; an element that is in both is counted only once). The other is the *intersection* of the two sets, written $E \cap F$ (sometimes called their product, and written $E \cdot F$ or EF); it consists of all elements that are in both E and F. If $E \cap F$ is empty, E and F are called *disjoint*; that is, E and F are disjoint if they have no element in common.

Exercise 1.2. Let Ω consist of the 26 letters of the alphabet. Let E consist of all the consonants (including y), and F of all the letters that occur in the words *real functions* (the n is counted only once). Show that (a) $E \cup F = \Omega$; (b) $F \supset C(E)$; (c) $C(F) \subset E$; (d) $F \cap E$ and $C(E)$ are disjoint.

There are various logical difficulties inherent in the uncritical use of the terminology of the theory of sets, and they have given rise to a great deal of discussion. Fortunately, however, they arise

---

[†]For simplicity of notation we frequently use, as here, a letter that has just been used as the name of a set, that we are now through with, to denote a different set.

only at a higher level of abstraction than we shall attain in the rest of this book, and in contexts that we should consider rather artificial, so that we may safely ignore them hereafter. Some forms of words which appear to define sets may actually not do so, somewhat as some combinations of letters which might well represent English words (e.g., "frong") do not actually do so. For example, although we can safely speak of sets whose elements are sets, we cannot safely talk about the set of all sets whatsoever. Supposing that we could, the set of all sets would necessarily have itself as one of its elements. This is a peculiar property, although there are other ostensible sets that have it, for example, the set of all objects definable in fewer than thirteen words (since this "set" is itself defined in fewer than thirteen words). We might well decide to exclude from consideration those sets that are elements of themselves. The remaining sets do not have themselves as elements; form the aggregate of all such acceptable sets, say A. Now is A one of the sets that we accept, or one of the sets that we exclude? If we accept A, it does not have itself as an element and so must be included in the aggregate of all sets with this property; that is, A belongs to A, and therefore we do not accept A. On the other hand, if we do not accept A, A is an element of itself; then since all elements of A are sets that are not elements of themselves, and so are acceptable, we must accept A. Thus if A is a set at all, we are involved in a logical contradiction. The only way out seems to be to declare that the words that seem to define A do not actually define a set.

Another paradoxical property of "the set of all sets" will turn up in §3.

Exercise 1.2a. A librarian proposes to compile a bibliography listing those, and only those, bibliographies that do not list themselves. Comment on this proposal.

2. Sets of real numbers. Since we have to start some- where, the reader will be supposed to be familiar with the real number system. Its algebraic properties—those con- nected with addition, subtraction, multiplication, and division, and with inequalities—will be taken completely for granted. However, there is one property of the real numbers that is less familiar to most people, even though it underlies concepts, such as limit and convergence, which are fundamental in calculus. This property can be stated in many equivalent forms, and the particular one that we select is a matter of taste. I shall take as fundamental the so-called *least upper bound property*. Before we can state what this property is, we need some more terminology. Let E be a nonempty set of real numbers. We say that E is *bounded above* if there is a number M such that every x in E satisfies the inequality $x \leqslant M$. For example, the set of all real numbers less than 2 is bounded above, and we can take $M = 2$, or $M = \pi$, or $M = 100$. On the other hand, the set of all positive integers is not bounded above. If E is bounded above, its *least upper bound* is B if B is the smallest M that can be used in the preceding definition. In our example, where E is the set of all real numbers less than 2, the least upper bound of E is 2. Another way of stating the definition of the least upper bound of E is to say that it is a number B such that every x in E satisfies $x \leqslant B$, while if $A < B$ there is at least one x in E satisfying $x > A$. The least upper bound of E may or may not belong to E. In the example just given, it does not. However, if we change the example so that E consists of all numbers not greater than 2, the least upper bound of E is still 2, and now it belongs to E.

So far, although we have talked about the least upper bound of a set, we have not known (except in our illustrative examples) whether there is any such thing. The least upper bound property, which we take as one of the axioms about real numbers, is just that *every nonempty set E that is bounded above does in fact have a least upper bound*. In other words, if we form the collection of all upper bounds of E, this collection has a smallest element (hence the name). We denote the least upper bound of E by $\sup E$ or $\sup_{x \in E} x$ (sup stands for supremum). When $\sup E$ belongs to E we sometimes write $\max E$ instead. Thus $\max E$ is the largest element of E if E has a largest element. The greatest lower bound, denoted by inf, is defined similarly. (Cf. Exercise 2.2.)

An *interval* is a set consisting of all the real numbers between two other numbers, or of all the real numbers on one side or the other of a given number. More precisely, an interval consists of all real numbers x that satisfy an inequality of one of the forms $a < x < b$, $a \le x < b$, $a < x \le b$, $a \le x \le b$ (where $a < b$), $x > a$, $x \ge a$, $x < a$, or $x \le a$. Using a square bracket to suggest $\le$ or $\ge$ and a parenthesis to suggest $<$ or $>$, we shall often use the following notations for the corresponding intervals: (a,b), $[a,b)$, $(a,b]$, $[a,b]$, (a,∞), $[a,\infty)$, $(-\infty,a)$, $(-\infty,a]$. Thus $(0,1]$ means the set of all real numbers x such that $0 < x \le 1$. (The use of the symbol ∞ in the notation for intervals is simply a matter of convenience and is not to be taken as suggesting that there is a number ∞.)

Exercise 2.1. For each of the sets E described below, describe the set of all upper bounds, the set of all lower bounds, $\sup E$, and $\inf E$.

(a) E is the interval $(0,1)$.
(b) E is the interval $[0,1)$.
(c) E is the interval $[0,1]$.
(d) E is the interval $(0,1]$.
(e) E consists of the numbers $1, \frac{1}{2}, \frac{1}{3}, \frac{1}{4}, \ldots$.
(f) E is the set containing the single point 0.
(g) $E = \{n + (-1)^n\}$, $n = 1, 2, \ldots$.
(h) $E = \{n + (-1)^n/n\}$, $n = 1, 2, \ldots$.

(i) E is the set of real numbers x, $0 < x < \pi$, such that $\sin x > \frac{1}{2}$.

Exercise 2.2. Give a detailed definition of $\inf E$, formulate a greatest lower bound property, and prove that it is equivalent to the least upper bound property.

If E is not bounded above, we write $\sup E = +\infty$; if E is not bounded below, we write $\inf E = -\infty$. These are convenient abbreviations, but are not to be interpreted as implying that there are real numbers $+\infty$ and $-\infty$; there are not. We can, if we like, create such infinite numbers and adjoin them to the real number system, but for most purposes it is undesirable to do so. No matter how we introduce infinite numbers, we are bound to make arithmetic worse than it already is: there is one impossible operation to begin with (division by zero), but if we make this operation possible we introduce even more impossible operations.

Exercise 2.3. Explore the consequences of introducing numbers $+\infty$ and $-\infty$ such that $a/0 = +\infty$ if $a > 0$, $a/0 = -\infty$ if $a < 0$. Can a reasonable meaning be given to $+\infty + (-\infty)$? To $0 \cdot (+\infty)$?

If a set is both bounded above and bounded below, it is called *bounded*. A bounded nonempty set E is characterized by having $\inf E$ and $\sup E$ both finite, or equivalently by being contained in some finite interval (a, b).

We have supposed in our discussion of upper and lower bounds that we have been considering nonempty sets. So that we shall not have to make reservations about whether or not our sets have elements, we make the (rather odd) convention that if E is empty, $\sup E = -\infty$ and $\inf E =$

$+\infty$. This convention allows us to say, for example, that $\sup(E \cup F)$ is the larger of $\sup E$ and $\sup F$, without having to consider whether E or F may be empty.

Exercise 2.4.* If E is not empty, $\inf E \leqslant \sup E$; there is strict inequality if E contains at least two points.

3. Countable and uncountable sets. If we have a set E with (say) five elements, for instance, the fingers on a hand, we can count these elements (or, for short, count E). This means exactly what we might expect: we can point to the elements of E, one by one, naming the integers 1, 2, 3, 4, 5, successively, as we do so. In slightly more formal language, we label all the elements of E by using the integers 1, 2, 3, 4, 5 once each. In still more formal language, we put the elements of E into *one-to-one correspondence* with the integers 1, 2, 3, 4, 5.

It turns out to be useful to extend the concept of counting to sets that have an infinite number of elements. Suppose, for example, that E is now the set of all even positive integers. We can no longer point to *all* the elements of E, one after another, naming successive integers, because E has too many elements. However, we can still imagine labeling all the elements of E with all the positive integers, so that each element of E has a different integer attached to it: we just have to label each even integer with the integer whose double it is, that is, we label the integer $2n$ with the label n. Since the even integers can be put into one-to-one correspondence with all the integers in this way, we may reasonably say that there are the same number of each, without, however, committing ourselves as to what the number of integers may be.

---

*An exercise that simply makes a statement calls for a proof of that statement.

The interest of this notion of counting comes largely from the fact that there actually are sets that cannot be counted, as we shall prove soon, so that we can classify sets according to whether they can be counted or not. The ones that cannot be counted can be thought of as "bigger" than those that can be counted. (As we have just seen, a set is not necessarily bigger, in this sense, than one of its proper subsets.) Before establishing the existence of uncountable sets, we shall introduce some more terminology and give additional examples of sets that can be counted.

To count a set means to put its elements into one-to-one correspondence with some set of consecutive positive integers starting from 1; this set is not necessarily to be a proper subset of the set of all positive integers. If a set can be counted we call it *countable*. (The empty set is also called countable: here the relevant subset of the positive integers can be thought of, with a little effort, as the empty set.) We can write the elements of a nonempty countable set in some such form as $x_1, x_2, x_3, \ldots$, where a typical element of the set would be denoted by x and the subscripts are the consecutive integers that are used as labels in counting the set. If we start out counting a countable set, either eventually we find a last element or else the counting process continues indefinitely. In the first case, the set is called *finite*; in the second, *countably infinite*.

The set of all the positive and negative integers together is countable, since we can use the odd positive integers to label all the positive integers and the even positive integers to label all the negative integers, thus:

elements	$\ldots$	-3	-2	-1	1	2	3	$\ldots$
labels	$\ldots$	6	4	2	1	3	5	$\ldots$

Exercise 3.1. Show similarly that *the union of any two countably infinite sets is countable.*

A less obviously countable set is the set of lattice points in the plane: these are the points with both coordinates integral, for example $(1, 2)$ or $(-5, 18)$. It is easy to see how to count them from the following diagram (which, for

$\cdots \cdots$

$(0, 2)$ $(1, 2)$ $(2, 2)$ $\cdots$

$(0, 1)$ $(1, 1)$ $(2, 1)$ $\cdots$

$(0, 0)$ $(1, 0)$ $(2, 0)$ $\cdots$

convenience, shows only the lattice points in the first quadrant; we can count all the lattice points by using Exercise 3.1 repeatedly). Without drawing the picture, we can think of first grouping all the lattice points (m, n) for which $m + n$ is $0, 1, 2, 3, \ldots$, and then counting the groups, one after another. What we have done in the diagram is to represent the lattice points in the first quadrant as the union of a countable collection of countable sets: the sets are the successive horizontal rows.

Some rather more complicated sets can be discussed by using the fact that *every subset of a countable set is countable.* Expressed in another way, this theorem says that a set whose elements can be labeled with some of the positive integers, using each only once, can equally well be labeled with all the positive integers. To see this, observe that each element of the given subset of a countable set is labeled with some positive integer. Take the one with the smallest label and relabel it 1; then take the element with the smallest label from the rest of the subset (if there is any more of the subset) and relabel it 2; and so on.

It is now easy to see that *the positive rational numbers form a countable set*. A positive rational number can be represented as a fraction p/q in lowest terms, where p and q are positive integers. If we associate the fraction $3/11$ with the lattice point $(3, 11)$, and generally p/q with (p, q), we have the rational numbers in one-to-one correspondence with a subset of the lattice points, that is, with a subset of a countable set. Hence the positive rational numbers form a countable set.

Exercise 3.2. Show that *the union of any countable collection of countable sets is countable*.

Exercise 3.2a. The set of all points inside a circle is called a disk. Let S be a set of nonoverlapping disks in the plane—that is, no disk in S has a nonempty intersection with any other disk in S. Show that S must be countable.

A still less obvious example is furnished by the algebraic numbers: these are the numbers (real or complex) that can be roots of polynomials with integral coefficients (for example, all rational numbers, $\sqrt{2}$, i, $2\sqrt[3]{3} + \sqrt[5]{7}$). To see that *the set of algebraic numbers is countable*, we notice first that there are only countably many linear polynomials with integral coefficients, countably many quadratic polynomials with integral coefficients, and so on.

Exercise 3.3. Why is this?

The polynomials of a given degree n with integral coefficients have at most n roots each, thus a countable number altogether. The aggregate of roots of polynomials of every degree with integral coefficients is accordingly a countable collection of countable sets, and so countable.

A more abstract example is the class of all finite subsets of a given countable set. For, the class of subsets with one element each is countable, the class of subsets with two elements each is countable, and so on. We again have a countable collection of countable sets. (As we shall see later (p. 17), the set of *all* subsets of a countably infinite set is not countable.)

The notion of countability can sometimes be used to show the existence of things of a particular kind. For a simple illustration, we prove that *not all real numbers are algebraic*. (A number that is not algebraic is called transcendental.) The real algebraic numbers are, as we know, countable; our first step is to suppose that they have been counted, and represented as decimals. It will simplify the notation somewhat, and do no harm, if we consider only real numbers between 0 and 1.

Every real number between 0 and 1 has an ordinary decimal expansion, for example,

$$\tfrac{1}{7} = 0.\overline{142857}1428571428 \ldots ,$$

$$\pi - 3 = 0.14159265358979323846 \ldots .$$

Conversely, every such expansion defines a real number between 0 and 1; if we write down, for example,

$$x = 0.123456789101112131415 \ldots ,$$

x is certainly a real number between 0 and 1, although we cannot connect it in any simple way with more familiar numbers. (For more details about decimals see §6.)

Now suppose that the real algebraic numbers between 0 and 1 have been counted, so that there is a first, a second, a third, and so on; call them a_1, a_2, a_3, and so on. We then have a column of decimals, which might start like this:

$$a_1 = 0.215367 \ldots$$
$$a_2 = 0.652489 \ldots$$
$$a_3 = 0.061259 \ldots$$
$$a_4 = 0.300921 \ldots$$

$$\ldots \quad \ldots$$

This list is supposed to contain *all* the real algebraic numbers between 0 and 1; that is, any such number will appear in the list if we go far enough. It is now easy to construct a decimal that does not appear anywhere in this list and so cannot be an algebraic number. For example, if we write down 0.5655 we have the start of a decimal that differs from a_1 in the first decimal place, from a_2 in the second, from a_3 in the third, and from a_4 in the fourth; obviously it is not going to be any of these four numbers. We can go on in the same way, putting 5 in the nth decimal place if a_n has anything except 5 there, and putting 6 in the nth place if a_n does have 5 there. The resulting decimal differs from each a_n in the nth decimal place and so cannot appear in our hypothetical list of all algebraic numbers, so it is not algebraic.

We can describe the construction more concisely if we let the digits of the nth algebraic number be $a_{n,1}, a_{n,2}, a_{n,3}, \ldots$, and construct a new number $b = 0.b_1 b_2 b_3 \ldots$ by taking $b_n = 5$ if $a_{n,n} \neq 5$, and $b_n = 6$ if $a_{n,n} = 5$. (There is no particular significance to the numbers 5 and 6.)

It is sometimes alleged that a proof of this kind is only a "pure existence proof" and furnishes no explicit example of a transcendental number. This is not the case. At least in principle, it is possible to count the algebraic numbers explicitly, find their decimal expansions, and so write down as many digits as we like of at least one transcendental number. The reason that the number π, say, seems more concrete is that π occurs in more contexts than the number we have just been talking about, so that more is known about it; in particular, people have been interested enough to compute many thousands of decimal places of π already.[1]*

We have shown how to find a transcendental number; to show that some given number is transcendental is much harder: it is a problem in number theory and requires deeper methods than the simple argument used here. The transcendence of e is fairly difficult, that of π considerably harder, e^π and $2^{\sqrt{2}}$ much harder again; and it is not known whether π^e is transcendental or not, or even whether it is irrational. Another transcendental number is $0.10100100000010\ldots$, where there are $n!$ zeros after the nth 1. This transcendental number is in a sense simpler than π or e, since we could say without much trouble what any particular digit, say the billionth, is, whereas we cannot, at least at present, do this for π or e.

If we look over the proof of the existence of transcendental numbers, we see that no use was made of the fact that $a_1, a_2, \ldots$ were algebraic numbers beyond the fact that the algebraic numbers form a countable set. The same argument applies, word for word, to show that if E is *any* given countable set of real numbers between 0 and 1, there is a real number between 0 and 1 that is not in E. Hence no countable set can exhaust the set of real numbers between 0 and 1, or in other words *the set of real numbers between* 0 *and* 1 *cannot be countable*.

---

*Superscripts in the text refer to the notes at the back of the book.

Exercise 3.4. The particular interval $(0, 1)$ is not significant; modify the preceding argument, or use the result, to show that the set of real numbers in any interval, however short, is not countable.

Exercise 3.5. Show that the set of real numbers in $(0, 1)$ whose decimal expansions do not contain any 3's is not countable. (The number 3 has no particular significance here.)

Exercise 3.5a. Criticize the following "proof" that the set of real numbers between 0 and 1 is countable: First count the decimals that have only one nonzero digit; then those with at most two nonzero digits, and so on; we have then broken the set into a countable set of countable sets.

We defined a set to be finite if it is countable but not countably infinite. We naturally call a set *infinite*, whether it is countable or not, as long as it is not finite. *Every infinite set contains a countably infinite subset.* To see this, choose a first element x_1, quite arbitrarily. The set with x_1 removed is still infinite (why?); choose an element x_2 from this reduced set; and so on. This process cannot terminate (again, why?), so our set contains the countably infinite subset $x_1, x_2, \ldots$.

Exercise 3.6. Supply answers to the two questions in the preceding paragraph.

Exercise 3.7. Show that if E is any infinite set and F is E with one point deleted, E and F can be put into one-to-one correspondence with each other. Thus *any infinite set can be put into one-to-one correspondence with a proper subset of itself.*

Exercise 3.8. Establish a one-to-one correspondence between a finite interval and the set of all real numbers.

As another application of the kind of reasoning that we have been using, we prove that *the aggregate A of subsets of any given nonempty set E is "larger" than E, in the sense that A cannot be put into one-to-one correspondence with E, or indeed with any subset H of E.* We shall not make any use of this fact, but it helps to justify the remark on p. 4 about the paradoxical character of "the set of all sets."

Exercise 3.9. Use the theorem just stated to show that the notion of the set of all sets *is* paradoxical.

For finite sets we could show without much trouble that there are two subsets of a set with one element (the set itself and the empty set); four subsets of a set with two elements (the whole set, two sets containing one element each, and the empty set); eight subsets of a set with three elements; and generally 2^n subsets of a set with n elements. So our statement is true for finite sets; the general proof will, in fact, cover all sets with at least one element.

Suppose, then, that E is a set with at least one element, and suppose that the collection of all the subsets of E can be put into one-to-one correspondence with a subset H of E. In other words, suppose that the subsets can be labeled, say as F_x, where x runs through the elements of H, so that every subset is labeled and no element of H is used twice. We are now going to deduce a contradiction, and this contradiction will show that the alleged one-to-one correspondence cannot exist. We form a subset G of E in the following way. For each x in H, we look at F_x and see whether F_x contains x. If F_x does not contain x, put x in G. (In particular, the x for which F_z is empty is put in G; the x for which F_x is E is not put in G.) Then G is a proper subset of E, and so by assumption corresponds to some z in H, that is, G is F_z. However, by construction, if z is in F_z, we did not put the element z into G, so G is not F_z; if, on the other hand, z is not in F_z, G contains z and F_z does not, so again G is not F_z. We have thus deduced from our initial assumption the contradictory statements that G is F_z and that G is not F_z, so that the initial assumption is untenable.

The preceding theorem shows, for example, that *there are more sets of real numbers than there are real numbers*.

In a rather similar way we could show that there are more real-valued functions, with domain the real numbers, than there are real numbers.

Exercise 3.9a.   Prove the preceding statement.

We now establish the fact, which seems surprising at first sight, that there are just as many points in a line segment as in a square area: that is, *the real numbers between 0 and 1 can be put into one-to-one correspondence with the points in a square*. (The points in a square are ordered pairs of real numbers, their coordinates; cf. p. 21.) The general idea of the correspondence is easy to grasp: if we have two real numbers, represented as decimals, we can interlace their digits to get a single real number; conversely, given a real number, we can dissect its decimal expansion to get a pair of real numbers. The details are not quite obvious, however. Suppose, to make decimals unique, that we select the nonterminating decimal when there is a choice: thus we take $0.243999\ldots$ instead of $0.244000\ldots$. The obvious procedure of making (p, q) with $p = 0.p_1 p_2 p_3 \ldots$, $q = 0.q_1 q_2 q_3 \ldots$ correspond to $0.p_1 q_1 p_2 q_2 \ldots$ does not work because, for example, the decimal $0.13201020\ldots$ would correspond to (p, q) with $p = 0.1212\ldots$, $q = 0.300\ldots$, and the latter is a decimal of the kind we are not allowing. Once we recognize this difficulty, we can easily avoid it, however. All that is necessary is to attach to each nonzero digit any string of consecutive zeros that immediately precedes it, and treat these groups of digits as units. Thus 0.13201020 . . . now corresponds to (p, q) with $p = 0.1202 \ldots$, $q = 0.301 \ldots$; and to (p, q) with $p = 0.003100054 \ldots$, $q = 0.100359 \ldots$ corresponds the real number $0.003110030005549 \ldots$.

It is often quite hard to exhibit a one-to-one correspondence between two sets explicitly; it is sometimes easier to show that each set can be put into one-to-one correspondence with a subset of the other. The following proposition, known as the Schroeder-Bernstein theorem, is useful in such situations. *If A and B are sets, if A can be put into one-to-one correspondence with a subset of B, and if B can be put into one-to-one correspondence with a subset of A, then A and B can be put into one-to-one correspondence with each other.*[1a]

We may suppose that initially the subsets of B and A with which we are concerned are not B and A themselves, since if they are, there is nothing to prove. We are supposed to have two one-to-one correspondences, one (call it S) between A and a subset of B, the other (call it T) between B and a subset of A. Take any element a_1 of A, find its image b_1 in B under S, find the image a_2 of b_1 under T, and so on. This process may lead us back to a_1 after a finite number of steps, or it may go on indefinitely. What it cannot do is to give a chain of elements that crosses itself, for example, so that $a_5 = a_2$; for, if this happened, T would carry b_1 into a_2 and also carry b_4 into a_2. This would contradict the assumption that T is one-to-one unless $b_1 = b_4$, in which case $a_1 = a_4$. In addition, it may happen that a_1 occurs as the image of some element of B under T, and in this case we can prolong the chain backwards from a_1, possibly indefinitely. If any elements of A remain, pick one and start a new chain.

In this way, the elements of A fall into disjoint classes: A_1 consists of elements that belong to chains with one pair of elements (symbolically, $a_1 \overset{S}{\to} b_1 \overset{T}{\to} a_1$); A_2 consists of elements that belong to chains with two pairs of elements ($a_1 \overset{S}{\to} b_1 \overset{T}{\to} a_2 \overset{S}{\to} b_2 \overset{T}{\to} a_1$); and so on. There may be, in addition, elements of

A that belong to infinite chains. There are three kinds of infinite chains: those with a first element in A that has no antecedent in B to produce it under T; those with a first element in B; and those with no first element at all. We call these classes A_0, A_{-1}, A_∞, respectively. The classes $A_{-1}, A_0, A_1, A_2, \ldots, A_\infty$ are all disjoint from each other and every element of A is in one of them. Let B_k be the class of elements of B that belong to chains containing elements of A_k. Then the B_k are also disjoint and every element of B is in one of them (since we can start a chain from an element of B as well as from an element of A).

Now A_1 and B_1 are obviously in one-to-one correspondence already. A_2 consists of pairs of elements of A connected with pairs of elements of B_2; we put A_2 and B_2 into one-to-one correspondence by pairing off the elements in the obvious way (the first element of a pair in A with the first element of the corresponding pair in B, and so on). We proceed similarly with A_k, B_k for $k = 3, 4, \ldots$. We put A_0 into one-to-one correspondence with B_0 by operating on each chain separately: pair the first element a_1 with its image b_1 under S; pair a_2 with b_2; and so on. In other words, use S on A_0 to carry A_0 into B_0. Here we use the fact that chains of elements in A_0 do not terminate. Similarly, for A_{-1} we use T to establish the one-to-one correspondence. Finally, A_∞ and B_∞ consist of chains that are infinite in both directions; here we can use either S or T. Thus we have established one-to-one correspondences between each A_k and the corresponding B_k, and hence a one-to-one correspondence between all of A and all of B.

As an application of the Schroeder-Bernstein theorem we show that *there are just as many sets of positive integers as there are real numbers* (in contrast to the fact, noted on p. 12, that there are only countably many *finite* sets of positive integers). In the first place, if we have a real number r between 0 and 1, we can represent it as a nonterminating decimal, for instance, as $0.20015907 \ldots$. To this real number we assign the set of integers 20, 10000, 500000, 9000000, \ldots . In the general case, if the digit $a \neq 0$ occurs in the nth decimal place of r, we incorporate in our set the integer whose decimal representation is a, followed by n

zeros. In this way the set consists of different integers, and two different real numbers r generate different sets of integers.

It may seem at first sight that we get only a relatively small proportion of all the possible sets of integers in this way. However, let us now consider an arbitrary set S of positive integers. To S we assign a unique real number as follows. First write the decimal $u = 0.123456789101112\ldots$ (formed by writing down all the positive integers in their natural order). If the integer n occurs in S, replace it in u by a string of zeros. For example, if $S = \{1, 8, 12, 13, 17\}$, the corresponding decimal will be

$$0.02345670910110000141516001819\ldots.$$

If S consists of all even positive integers, its decimal representative is $0.10305070900110013\ldots$. Thus we have a one-to-one correspondence between all sets of positive integers and a set of real numbers, and another one-to-one correspondence between the set of all real numbers and a class of sets of positive integers. Then, by the Schroeder-Bernstein theorem, there is a one-to-one correspondence between the class of all sets of positive integers and the class of all real numbers.

Exercise 3.10. Show that there are just as many sequences of real numbers as there are real numbers.

4. Metric spaces. A *space* is just another name for a set, with emphasis on the possibility of considering its subsets. However, when we call a set a space we usually intend to imply that some sort of additional conditions are to be imposed on the points of the set, which, of course, need not be points in the ordinary sense. A *metric space* is a (nonempty) set in which we can speak of the distance between two points. It is a generalization of the ordinary lines, planes, and three-dimensional spaces of geometry, where, in making the generalization, only some of the geometrical properties have been preserved.

We require the distance between two points to satisfy the following conditions (which the ordinary distance in Euclidean geometry certainly does satisfy): the distance is a nonnegative real number, zero only if the points coincide; it is the same in either direction; and the sum of two sides of a triangle is at least as much as the third side. In symbols, if $d(x, y)$ denotes the distance between the points x and y, we are to have

(1) $d(x, y) \geqslant 0$; $d(x, x) = 0$; $d(x, y) > 0$ if $x \neq y$ (positivity);

(2) $d(x, y) = d(y, x)$ (symmetry);

(3) $d(x, z) \leqslant d(x, y) + d(y, z)$ (triangle inequality).

We often refer to the distance function for the space as the *metric* of the space.

It turns out that a good deal of geometry depends only on these three properties of distance. Consequently, many facts about ordinary space can be carried over to other spaces that at first sight are very different because their points are not points in the ordinary sense, but may, for example, be functions. The possibility of using geometrical language in metric spaces makes many of their properties more intuitive, although, of course, it may on occasion be misleading as well.

Here are some examples of metric spaces.

R_1, Euclidean one-dimensional space, is the set of all real numbers, with $d(x, y) = |x - y|$.

R_2, Euclidean two-dimensional space, is the ordinary plane of analytic geometry and the distance is the ordinary distance. Its points are ordered pairs of real numbers ("ordered" means that (x, y) is not the same as (y, x)).

The distance from (x_1, y_1) to (x_2, y_2) is

$$\left\{ (x_1 - x_2)^2 + (y_1 - y_2)^2 \right\}^{1/2}.$$

R_n, Euclidean n-dimensional space, is defined similarly.

Exercise 4.0. Are the following objects metric spaces, or not?

(a) The Euclidean plane with points $x = (x_1, x_2)$, etc., but with distance defined by $d(x, y) = |x_1 - y_1|$.

(b) The set of cities in the United States that have direct airline service, with $d(A, B) =$ "scheduled air travel time between A and B."

(c) The real positive numbers with $d(x, y) = x/y$.

(d) As in (c) but with $d(x, y) = |\log(x/y)|$.

(e) The set of all positive numbers represented by terminating decimals, with $d(x, y) = |x - y|$ rounded to 10 decimal places.

Exercise 4.1. If we use the same points but change the definition of distance we get a new space. For instance, change the distance in R_2 by saying that the distance from (x_1, y_1) to (x_2, y_2) is $|x_1 - x_2| + |y_1 - y_2|$; show that the result is a metric space; show that there is now a nondegenerate triangle such that the sum of two sides is equal to the third side; draw the locus of points that are at unit distance from $(0, 0)$. Do the same things if the distance from (x_1, y_1) to (x_2, y_2) is taken to be the larger of $|x_1 - x_2|$ and $|y_1 - y_2|$.

In our next three examples the elements of the space will be infinite sequences of numbers. Since the elements of R_n are sequences of n numbers, these "sequence spaces" may be thought of as infinite-dimensional generalizations of R_n.

c_0 is the space of sequences that converge to zero. Its points are sequences of numbers: $\{x_1, x_2, x_3, \dots\}$, where $\lim x_n = 0$. If we denote such a sequence by the single letter x, the distance $d(x, y)$ is defined to be $\sup_{n \geqslant 1} |x_n - y_n|$.

For example, if $x = \{1, \frac{1}{2}, -\frac{1}{3}, -\frac{1}{4}, \frac{1}{5}, \dots\}$ and $y = \{1, -1, 0, 0, \dots\}$, then $d(x, y) = \frac{3}{2}$.

m is the space of bounded sequences. Its elements are again sequences of numbers, but now required only to be bounded. The distance is the same as for c_0. Thus we could have $x = \{1, 0, 1, 0, \dots\}$ and

$$y = \{1, 4, 1, 5, 9, 2, 6, 5, \dots\},$$

where no y_n is to be negative or exceed 9. Here the distance $d(x, y)$ cannot be computed until we know more about the law of formation of y. Given that $y_9 = 3$, $y_{10} = 5$, $y_{11} = 8$, and $y_{12} = 9$, however, we see that $d(x, y) = 9$, the largest possible value.

l^2 is the space of sequences for which the sum of the squares of the components converges. Its elements are sequences $x = \{x_1, x_2, \dots\}$ with $x_1^2 + x_2^2 + x_3^2 + \cdots < \infty$. We take $d(x, y) = \{\sum_{n=1}^{\infty}(x_n - y_n)^2\}^{1/2}$. To verify the triangle inequality in this case, we need Minkowski's inequality (see p. 176):

$$\begin{aligned}
d(x, z) &= \left\{ \sum (x_n - z_n)^2 \right\}^{1/2} \\
&= \left[\sum \{(x_n - y_n) + (y_n - z_n)\}^2 \right]^{1/2} \\
&\leqslant \left\{ \sum (x_n - y_n)^2 \right\}^{1/2} + \left\{ \sum (y_n - z_n)^2 \right\}^{1/2} \\
&= d(x, y) + d(y, z).
\end{aligned}$$

Next we have some examples of metric spaces whose points are functions.

C is the space of continuous functions defined on the closed interval $[0, 1]$. Its elements are continuous functions

$x = x(t), 0 \leqslant t \leqslant 1$, and

$$d(x, y) = \max_{0 \leqslant t \leqslant 1} |x(t) - y(t)|.$$

For example, we could have $x(t) = \cos \pi t$ and $y(t) = 2t - 1$, and then $d(x, y) = 2$.

B is the space of bounded functions defined on $(0, 1)$, with $d(x, y) = \sup_{0 \leqslant t \leqslant 1} |x(t) - y(t)|$. (We have to write sup instead of max here since $|x(t) - y(t)|$ may fail to attain its maximum.) For example, $x(t)$ could be defined as $1 - n^{-1}$ in the interval $(1/(n + 1), 1/n)$, $n = 1, 2, \ldots$; if 0 stands for the function that takes the single value 0, then $d(x, 0) = 1$.

The space of continuous functions on some interval, say $[0, 1]$, forms a metric space with

$$d(x, y) = \left\{ \int_0^1 |x(t) - y(t)|^2 \right\}^{1/2}.$$

Exercise 4.2. Any (nonempty) subset of a metric space is again a metric space with the metric that it inherits from the original space.

Exercise 4.3. Any nonempty set whatsoever can be made into a metric space if we introduce a metric by $d(x, y) = 1$ if $x \neq y$, and $d(x, x) = 0$.

5. Open and closed sets. There are several special kinds of sets in metric spaces that occur so frequently that they need to have names. Sections 5, 6, and 7 are mainly devoted to introducing them and making them seem familiar.

A *neighborhood* of a point x is a generalization of a circular disk (interior of a circle) with center at x: it is the set of all points y that are distant less than some positive r

from the point x; in symbols, $d(x, y) < r$. In fact, in R_2 the neighborhoods of x are just circular disks centered at x. In R_1, they are intervals centered at x; in R_3, they are solid spheres. If $r < 1$, the neighborhoods in the space of exercise 4.3 are single points.

A set is called *bounded* if it is contained in some neighborhood. Thus the interval $(0, 1)$ in R_1 is bounded, whereas the interval $(1, \infty)$ is not. In R_1, this definition agrees with the one that we used earlier, that a set is bounded if it is both bounded above and bounded below. In R_1, a bounded set has a least upper bound and a greatest lower bound, but there is nothing corresponding to this property in general metric spaces.

Exercise 5.1. Describe a neighborhood in the space C.

Exercise 5.2. Describe the neighborhoods in the space consisting of the points of R_2 that have two integral coordinates, with the R_2 metric.

If E is a set in a metric space and x is a point of E, we say that x is an *interior* point of E if some neighborhood of x (possibly a small one) consists exclusively of points of E. The idea of the definition is to give a set an interior that corresponds fairly closely to the intuitive notion of "interior," and in a space like R_2 it succeeds fairly well. For example, the set in R_2 of points (x, y) such that $0 \leqslant x \leqslant 1$ and $0 \leqslant y \leqslant 1$ is a square, and its interior consists of all the points in the square that are not on its perimeter. On the other hand, the set of rational points of R_1 has no interior points at all. In exercise 4.3 we considered the space consisting of arbitrary points with $d(x, y) = 1$ or 0 according as $x \neq y$ or $x = y$. In this space every point is an interior point of every set that contains it.

Indeed, if any set in a metric space is regarded as a new metric space in itself, with the original metric, all its points become interior points of the new space. Thus the notion of interior point depends not only on the set we are considering, but also on the space in which the set lies.

Again let E be a set in a metric space; if x is not necessarily a point of E, but every neighborhood of x (with emphasis on its possible smallness) contains both at least one point of E (possibly only x itself) and at least one point of $C(E)$ (again, possibly only x itself), x is called a *boundary point* of E. The *boundary* of E means the set of all boundary points of E. For a square in R_2, the boundary is just what we might expect. In R_1, the boundary of the interval $[a,b]$ or of the interval (a,b) consists of the two points a and b; so does the boundary of the set consisting of the two points a and b.

The term *frontier point* is sometimes used instead of boundary point, and would perhaps be preferable: the idea of a boundary point has nothing to do with the idea of boundedness. An unbounded set can (and often does) have a nonempty boundary. For example, the interval $(0, \infty)$ in R_1 has the point 0 as its boundary; R_1, considered as a subset of R_2, has itself as boundary. On the other hand, a nonempty bounded set may have an empty boundary (although we shall see that this cannot happen in R_1 or R_2).

Exercise 5.2a. Describe the interior points and boundary points of each set:

(i) In R_2, the circumference $x^2 + y^2 = 1$.

(ii) In R_1, the union of the open intervals $(1/(n + 1), 1/n)$, $n = 1, 2, \ldots$.

(iii) In R_2, the union of the open rectangles of height 1 standing on the intervals of (ii).

(iv) The set in R_2 indicated in the picture on p. 31.

Exercise 5.3. For the space of exercise 5.2, show that the boundary of every set is empty.

Exercise 5.4. Show that E and $C(E)$ have the same boundary.

Exercise 5.5. If E is a set and B is the boundary of E, show that the boundary of B is a subset of B, and that it may be a proper subset.

Exercise 5.6. Let N be a neighborhood of x, of radius r. What can be said about the boundary of N (a) if the underlying space is R_2? (b) if it is an arbitrary metric space?

A set all of whose points are interior points is called *open*; a set that contains all its boundary points is called *closed*. As we shall see, a set may be neither open nor closed, and a set may be simultaneously open and closed. These notions depend on the space in which the set lies, as well as on the set itself.

Exercise 5.7. In R_1, the interval (a, b) is open (it is called an open interval for this reason), and the interval $[a, b]$ is closed (and called a closed interval).

Exercise 5.8. Are the intervals $[a, b]$ and (a, b) open, closed, or neither if considered as subsets of R_2?

Exercise 5.9. Show that the interval $[0, 1)$ is neither open nor closed in R_1.

Exercise 5.10. Show that the empty set and the whole space are always both open and closed.

Exercise 5.11. Consider the metric space consisting of the intervals $(n, n + \frac{1}{2})$ in R_1, $n = 0, \pm 1, \pm 2, \ldots$, with the R_1 metric. Show that this space has many sets that are both open and closed.

Exercise 5.12. Is the set of all rational points in R_1 open, closed, or neither?

Exercise 5.13. If the set of exercise 5.12 is considered as a space by itself, with the R_1 metric, show that it has many sets that are both open and closed.

Exercise 5.14. Show that all sets in the space of exercise 5.2 are both open and closed.

Exercise 5.15. Show that E is open if every point of E is contained in an open subset of E.

There are several alternative definitions of open set and closed set.

Exercise 5.16. *A set is open if and only if it contains none of its boundary points.*

Exercise 5.17. *A set is open if and only if its complement is closed.*

Exercise 5.18. *A set is closed if and only if its complement is open.*

Exercise 5.19. Define a *limit point* of E as a point x (whether in E or not) such that every neighborhood of x (again, with emphasis on the possible smallness of the neighborhood) contains at least one point of E other than x. The more descriptive term *cluster point* is also used. *A set is closed if and only if it contains all its limit points.*

Exercise 5.20. Every neighborhood of a limit point of E contains an infinite number of points of E.

Exercise 5.21. The set of limit points of E is closed.

Exercise 5.21a. The set of boundary points of E is closed.

Exercise 5.22. Find the limit points of the following sets in R_1: (a) the interval $(0, 1)$; (b) the set consisting of $1, \frac{1}{2}, \frac{1}{3}, \frac{1}{4}, \ldots$; (c) the set of rational points in $(0, 1)$.

Exercise 5.22a. Find the limit points of the following sets in R_2:
  (i) The set indicated in the picture on p. 31.
  (ii) The set of all points with coordinates $(1/m, 1/n)$, $n = 1, 2, 3, \ldots$; $m = 1, 2, 3, \ldots$.
  (iii) The set of all points with polar coordinates $(r, 1/n)$, $0 \leqslant r \leqslant 1$; $n = 1, 2, \ldots$.

Exercise 5.23. If $E = A \cup B$, every limit point of E is either a limit point of A or a limit point of B.

If f is a continuous real-valued function on a real interval (see §§12, 13 for a formal definition), and c is a given real number, the set of points x for which $f(x) < c$ is open, and the sets where $f(x) = c$ or where $f(x) \leqslant c$ are closed (see p. 83).

Open sets in R_1 have an especially simple structure; they *consist of a countable number of disjoint open intervals.* The word countable is redundant here: any collection of disjoint open intervals in R_1 is countable, since each interval contains a rational number that is in no other interval, and so our collection of intervals is in one-to-one correspondence with a subset of the rational numbers.

To show that a given nonempty open set G in R_1 is a union of intervals is rather tedious in detail, but the idea of the proof is simple. Since G is open and not empty, it contains a point and then a neighborhood of that point. We let this neighborhood, which is an open interval, expand until it is as large as possible. If the enlarged interval does not exhaust G, pick a new point and a neighborhood of it in G, and repeat the process; and so on. No point of G can escape, since if one did we could proceed to enclose it in an interval as before.

To do this carefully, it is convenient to suppose that G is bounded; otherwise we can cover G by a countable number of open intervals and combine the results for the intersections of G with these intervals. If we show that there is a largest interval contained in G and containing a given point x, there is a largest open interval contained in G (there are only a finite number of length greater than 1, a finite number of length greater than $\frac{1}{2}$, and so on). If there is more than one largest interval, we can arrange them according to the magnitudes of their left-hand endpoints. Then we do the same for the next largest intervals, and so on. In this way we can see that the representation of G as a union of open intervals is unique.

It remains to show that there is a largest open interval that is a subset of G and contains a given point x. There is certainly some interval (a, b) (a neighborhood of x) that is in G. Let B be the least upper bound of numbers b such that the interval (x, b) is in G. Similarly, let A be the greatest lower bound of numbers a such that the interval (a, x) is in G. Since G is bounded, A and B will be finite. Then B is not a point of G, since if it were, G would contain a neighborhood of B and B would not be an upper bound of the set used in defining it. A similar argument applies to A. Hence the interval (A, B) is in G and cannot be enlarged without including points outside G.

If we look for subsets (other than the empty set and the whole space) of R_1 or R_2 that are both open and closed, a little experiment will convince us that there are none. That this conclusion is in fact correct will be proved shortly. The property of R_1 and R_2 (and generally of R_n) that allows only trivial sets to be both open and closed is called connectedness. We define this property first for open sets. An open set (in particular, the whole space) is *connected* if it cannot be represented as the union of two disjoint open sets, neither of which is empty. Thus, for example, in R_1, the union of the two open intervals $(0,\frac{1}{2})$ and $(\frac{1}{2},1)$ is not connected, since the two intervals are each open and not empty, and are disjoint.

More generally, a set E is connected if it cannot be covered by two open sets whose intersections with E are disjoint and not empty. This notion of connectedness is perhaps not completely in accord with intuition. We shall show presently that R_1 and R_2 are connected. The set of rational points in R_1 is not connected, since, for example, it is covered by the open sets defined by the inequalities $x > \sqrt{2}$ and $x < \sqrt{2}$. On the other hand, a set such as the one indicated in the figure, consisting of an oscillating

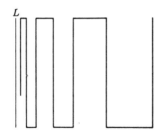

curve condensing toward a line segment, together with the line segment, *is* connected. (The graph of $y = \sin(1/x)$, together with the segment $-1 \leqslant y \leqslant 1$, has a similar character.) We might think that the line segment L at the left could be separated from the rest of the set, just as two abutting open intervals can be separated from each other. However, an open set that covers any point of L would have to contain a neighborhood of a point of L and the oscillating part of the graph enters any such neighborhood. Indeed, for the same reason, the set is still connected if we include only the rational points on L, or only the irrational points on L. By combining two sets of this kind it is possible to construct two connected sets inside a square, one of which joins two opposite corners of the square, and the other of which joins the other two corners, although the two sets have no point in common.

It is also possible to have a set that is connected but has the property that after one particular point is removed the remainder has no connected subsets containing more than one point.[2] (A construction is outlined on p. 41.)

A set may be open or not according to the space in which it is considered. However, we must not suppose, merely because connectedness was defined by using open sets, that a set can change from being connected to not being connected if it is considered as a subset of a different space. In fact, the property of being connected, unlike the property of being open, is an intrinsic property of the set. That is, if a set is connected when considered in one space, it is still connected when considered in another space, as long as the metric remains the same on points of the set.

We shall show (equivalently) that the property of not being connected depends only on the set. Suppose that E is a set in a metric space S, that E is not connected, and that S_1 is a subspace of S and S is a subspace of S_2, with $E \subset S_1$. We have to show that, whether we have added points to S (to get S_2) or taken points away from S (to get S_1), the set E is still disconnected.

We start with the assumption that $E \subset A \cup B$, where A and B are open (in S) and disjoint, and neither $A \cap E$ nor $B \cap E$ is empty.

The inference from S to the smaller space S_1 is easy. We replace the sets A and B by subsets A_1 and B_1 of S_1, where A_1 contains all the points of A that are in S_1, and B_1 contains all the points of B that are in S_1. These sets still cover E, have nonempty intersections with E, and are disjoint. Also, A_1 is open with respect to S_1, since a neighborhood in S_1 of a point p of A_1 consists of all points of S_1 whose distances from p are less than some r, and these points are all in A, since A is an open subset of S. Hence these points are in A_1. Similarly B_1 is open with respect to S_1. Hence E is still disconnected in S_1.

The inference from S to the larger space S_2 is harder. We take the covering sets A and B that show that E is disconnected in S, and inflate them into subsets A_2 and B_2 of S_2 as follows. Since A is open, if $p \in A$ the distances from p to points of B have a positive lower bound (the radius of a neighborhood of p that is in A). For each p in A we adjoin to A all the points of S_2 whose distances from p are less than half the greatest lower bound of the distances from p to points of B. We call the enlarged set A_2. Similarly we obtain B_2 by enlarging B.

The sets A_2 and B_2 still cover E and have nonempty intersections with E. Moreover, A_2 is open in S_2. For, if $q \in A_2$, there is a point p in A such that $d(p,q)$ is less than half the distance from p to any point of B. Then if q' is a point of S_2 close to q, it is also true that $d(p,q')$ is less than half the distance from p to points of B, so q' also belongs to A_2. That is, A_2 is open in S_2. Similarly B_2 is open in S_2.

Finally, A_2 and B_2 are disjoint. For, if $q \in A_2$ and $r \in B_2$, we obtain q and r from points $p \in A$ and $s \in B$; and

$$d(q,r) \geqslant d(p,s) - d(p,q) - d(s,r).$$

Since $d(p,q) < \frac{1}{2}d(p,s)$ and $d(s,r) < \frac{1}{2}d(s,p)$ by construction, we have $d(q,r) > 0$. This inequality implies that A_2 and B_2 are disjoint. We have therefore shown that E is disconnected in S_2.

It is easy to show that R_1 *is connected*. If it were not, it would be the union of two disjoint nonempty open sets. These sets would also be closed (Exercise 5.17), since each is the complement of the other. Hence it is enough to prove that R_1 has no nonempty subset that is both open and closed and is not all of R_1. Suppose there is such a subset; call it G. Then G is made up of a countable number of open intervals whose endpoints are not in G, as we have just shown. On the other hand, these endpoints are boundary points (also limit points) of G, and since G is also closed, they belong to G. This contradiction shows that G cannot exist.

To show that R_2 is connected it is sufficient to show that it contains no set, other than itself and the empty set, that is both open and closed. Let E be such a set; let $P \in E$ and $Q \in C(E)$. Consider the infinite straight line through P and Q, regarded as a space L, with the distance between two points in L equal to their distance in R_2. Then L is a copy of R_1. The sets $E \cap L$ and $C(E) \cap L$ are both open and closed in L, and neither of them is empty (since one contains P and the other contains Q). This contradicts the connectedness of R_1.

Exercise 5.24. Show that every nonempty set in R_2, except for R_2 itself, has a nonempty boundary.

Exercise 5.25. The *closure* of E is the union of E and the set of all limit points of E. Show that it is also the union of E and the set of all boundary points of E, and that it is closed.

Exercise 5.25a. If E is not finite, must some point of E be an interior point of the closure of E?

Exercise 5.26. What are the closures of the sets in Exercise 5.22?

Exercise 5.27. A neighborhood of x consists of the points y such that $d(x, y) < r$. Show that in R_1 or R_2 the closure of this neighborhood is the set of points y such that $d(x, y) \leqslant r$. Is this true in every metric space?

An important fact about closed sets is that *the union of two closed sets is still closed*, and *the intersection of two closed sets is still closed.* Since a closed set is characterized by containing all its limit points, to prove the first statement we consider two closed sets E_1 and E_2, and any limit point p of $E_1 \cup E_2$; we are to show that $p \in E_1 \cup E_2$. By Exercise 5.20, every neighborhood of p contains an infinite number of points of $E_1 \cup E_2$ other than p, hence either an infinite number of points of E_1 (other than p) or an infinite number of points of E_2 (other than p). This shows that p is a limit point of at least one of E_1 and E_2. Since E_1 and E_2 are both closed, $p \in E_1$ if p is a limit point of E_1, and $p \in E_2$ if p is a limit point of E_2. Hence p belongs to at least one of E_1 and E_2, that is, p belongs to $E_1 \cup E_2$. Therefore $E_1 \cup E_2$ contains all its limit points and so is closed.

To show that the intersection of two closed sets is closed, consider a limit point q of $E_1 \cap E_2$. Every neighborhood of q contains points, other than q, belonging both to E_1 and to E_2. This property makes q a limit point of E_1 and a limit point of E_2. Since E_1 and E_2 are both closed, q belongs to both and so to $E_1 \cap E_2$. Therefore $E_1 \cap E_2$ contains all its limit points and so is closed.

Exercise 5.28. Prove by induction that the union of any finite number of closed sets is closed and that the intersection of any finite number of closed sets is closed.

Exercise 5.29. Show that *the intersection of any number (finite or infinite) of closed sets is closed.*

Exercise 5.30. Find an example to show that the union of a countably infinite number of closed sets is not necessarily closed.

Exercise 5.31. By arguing in a similar way, or by considering the complements of the sets concerned, show that *unions of any number of open sets are open*; that intersections of finitely many open sets are open; and that intersections of a countable infinite number of open sets need not be open.

Exercise 5.32. If N_1 is a neighborhood consisting of all y such that, with a given x, $d(x, y) < r$, and N_2 is the neighborhood of the same x consisting of all y such that $d(x, y) < r/2$, show that the closure of N_2 is a subset of N_1. Must it be a proper subset?

A set E is called *perfect* if it is empty, or if it is closed and every point of E is a limit point of E. A closed interval in R_1 is perfect; so is the union of a finite number of closed intervals. In the next section we shall meet examples of more general perfect sets.

6. Dense and nowhere dense sets. A set E is *everywhere dense* (or, for short, just *dense*) if its closure is the whole space. In particular (in R_n, equivalently) E is dense when every point of the space is a limit point of E. A set is *nowhere dense* if its closure contains no neighborhoods. In other words, E is nowhere dense if E is empty, or if every neighborhood in the space contains a subneighborhood that is disjoint from E. The rational points in R_1 form a dense set. A set consisting of a finite number of points of R_1 is nowhere dense. We shall shortly meet some more complicated nowhere dense sets.

Note that "nowhere dense" is not the negative of "everywhere dense." A set that fails to be everywhere dense must have the property that its closure fails to fill some neighborhood (possibly small). If a set fails to be nowhere dense its closure must fill some neighborhood, but not necessarily the whole space.

Occasionally we need to say that a set E is dense in an interval, or in some other set, or is nowhere dense in an interval. Such phrases are self-explanatory.

Exercise 6.1. Consider the space Ω whose elements are the integral points $1, 2, 3, \ldots$ of R_1, with the R_1 metric. Describe the neighborhoods in Ω. Is the set containing the single point 1 a nowhere dense set in Ω?

Exercise 6.2. If a closed set contains no neighborhoods, it is nowhere dense.

Since every point of a perfect set is a limit point of the set, it would appear that a nonempty perfect set must have a great many points. It is therefore somewhat surprising to find that a nonempty set can be both nowhere dense and perfect.

Exercise 6.3. R_1, considered as a subset of R_2, is nowhere dense and perfect.

In R_1 there are no nowhere dense perfect sets of such a simple structure. An example of a nowhere dense perfect set in R_1 is the *Cantor set*, which may be used as the basis for constructing many examples of sets and functions with peculiar properties. This set is constructed as follows.

Consider the closed interval $[0, 1]$ in R_1. Remove the open middle third, that is, the interval $(\frac{1}{3}, \frac{2}{3})$. Next remove the open middle thirds of the two remaining intervals, that is, remove $(\frac{1}{9}, \frac{2}{9})$ and $(\frac{7}{9}, \frac{8}{9})$. Then remove the open middle thirds of the four remaining intervals; and so on indefinitely.

What remains? In the first place, what has been removed is a union of open sets (indeed, of open intervals), and so

is open; what remains is its complement (with respect to
[0, 1]), and so is a closed set. The endpoints of the various
middle thirds were not removed, so they remain; and since
the remaining set is closed, every limit point of endpoints
remains. For example, if we start from $\frac{1}{3}$ and take the
closest endpoint in the second step ($\frac{1}{3} - \frac{1}{9} = \frac{2}{9}$), then the
closest endpoint in the third step ($\frac{1}{3} - \frac{1}{9} + \frac{1}{27}$), and so on,
the (only) limit point of this set of points is
$\frac{1}{3} - \frac{1}{9} + \frac{1}{27} - \cdots = \frac{1}{4}$. Thus there are, in fact, limit points
of endpoints which are not endpoints. The Cantor set is
the set that remains after we have removed all the middle
thirds: it consists of all the endpoints and of their limit
points.

Exercise 6.3a. Does the Cantor set contain any irrational
points? If so, find one explicitly. Does the Cantor set contain the
point $\sqrt{\pi} - 1 = .77245 \ldots$?

We have just observed that the Cantor set is closed. It
also contains no interval, since the total length of intervals
removed is $\frac{1}{3} + \frac{2}{9} + \frac{4}{27} + \cdots = 1$. Therefore the Cantor
set is nowhere dense (Exercise 6.2). To show that it is
perfect we just have to show that each of its points is a
limit point. Since the limit points of endpoints are
naturally limit points of the set, it is merely a question of
showing that the endpoints are limit points. Consider, for
example, the point $\frac{1}{3}$. To the left of it there is an interval
of length $\frac{1}{3}$ from which we remove the middle third,
leaving an interval of length $\frac{1}{9}$ adjacent to the point $\frac{1}{3}$;
then we remove the interval ($\frac{1}{9}, \frac{2}{9}$), leaving an interval of
length $\frac{1}{27}$ adjacent to $\frac{1}{3}$; and so on. In any neighborhood
of $\frac{1}{3}$ there will always be a short interval that is not
removed at some step, and this interval will contain an

endpoint belonging to the next step. Hence $\frac{1}{3}$ is a limit point of endpoints. A similar argument applies to any other endpoint.

It will be useful later on to have a purely arithmetical construction for the Cantor set. We use the expansion of real numbers in "decimals" in bases 2 and 3 (binary and ternary expansions), instead of in base 10. For example,

$$0.10010110 \ldots \text{(base 2)}$$

means

$$\frac{1}{2} + \frac{0}{2^2} + \frac{0}{2^3} + \frac{1}{2^4} + \frac{0}{2^5} + \frac{1}{2^6} + \frac{1}{2^7} + \frac{0}{2^8} + \cdots,$$

while

$$0.10010110 \ldots \text{(base 3)}$$

means

$$\frac{1}{3} + \frac{0}{3^2} + \frac{0}{3^3} + \frac{1}{3^4} + \frac{0}{3^5} + \frac{1}{3^6} + \frac{1}{3^7} + \frac{0}{3^8} + \cdots.$$

(In base 2 we have only the digits 0 and 1, whereas in base 3 we have the digits 0, 1, and 2.) Thus $0.020202 \ldots$ (base 3) $= \frac{1}{4}$. (The reasoning is the same that is used in summing a repeating decimal in base 10: if $x = 0.0202 \ldots$, we have $9x = 2.0202 \ldots = 2 + x$.) Note that this is not the expansion of $\frac{1}{4}$ that we used above in showing that $\frac{1}{4}$ belongs to the Cantor set; but it is equivalent since

$$\frac{2}{3^2} + \frac{2}{3^4} + \cdots = \frac{3}{3^2} - \frac{1}{3^2} + \frac{3}{3^4} - \frac{1}{3^4} + \cdots$$

$$= \frac{1}{3} - \frac{1}{3^2} + \frac{1}{3^3} - \frac{1}{3^4} + \cdots.$$

Now let us express all the numbers between 0 and 1 in base 3. The numbers whose first digit is 1 are between $\frac{1}{3}$ and $\frac{2}{3}$ (inclusive), so they fill the first interval whose interior was discarded in forming the Cantor set. The numbers whose first digit is 0 fill the interval $[0, \frac{1}{3}]$, and the subset of these whose second digit is 1 are the numbers between $\frac{1}{9}$ and $\frac{2}{9}$, that is, they fill one of the intervals whose interior was excluded in the second step of the construction. Every number excluded so far has a 1 in the first or second place in its ternary expansion; so do the endpoints, but they also have expansions that have no 1's. For example, $\frac{1}{3} = 0.10000 \ldots = 0.02222 \ldots$, and $\frac{2}{3} = 0.12222 \ldots = 0.2000 \ldots$. By continuing in this way we see that the Cantor set can be described as consisting precisely of those numbers that have expansions in base 3 containing no 1's. The endpoints are the numbers of this kind whose expansions in base 3 end in all 0's or all 2's (which are equivalent, just as $0.1000 = 0.09999 \ldots$ in base 10).

Exercise 6.4.   Show that the Cantor set is uncountable.

Actually we can say more: the Cantor set can be put into one-to-one correspondence with the set of all real numbers between 0 and 1. Recall that the points of the Cantor set are just the numbers that can be written in base 3 using only 0's and 2's. With each such number x associate the number obtained by halving each digit in the ternary expansion of x and interpreting the result in base 2. In this way we get every number between 0 and 1 from some point of the Cantor set, and the endpoints of excluded intervals give rise to two different representations of the same number, for example, $\frac{1}{3} = 0.022 \ldots$ in base 3 and $\frac{2}{3} = 0.2000 \ldots$ in base 3 both yield $\frac{1}{2} = 0.0111 \ldots = 0.1000 \ldots$ in base 2. The correspondence is one-to-one between the Cantor set, less the countable set of endpoints of excluded intervals, and the set of all real numbers, less the countable set of numbers that have double representations in base 2. By making these countable sets correspond to each other we obtain a one-to-one correspondence between the Cantor set and the set of real numbers between 0 and 1.

We can use the Cantor set to construct the set mentioned on p. 32, which becomes totally disconnected after the removal of a single point. Let P be the point $(\frac{1}{2}, 1)$ and join P to the points of the Cantor set by straight line segments. Now delete the points with irrational ordinates on the lines going to endpoints of complementary intervals of the Cantor set, and delete the points with rational ordinates on the lines going to the other points of the Cantor set. The resulting set is connected but when P is removed it contains no connected subsets except for single points. It is known as the Cantor teepee.[2a]

A metric space is called *separable* if it contains a countable set that is everywhere dense. For example, R_1 is separable because the rational numbers form a countable dense set.

Exercise 6.5. Show that R_2 is separable.

The space c_0 (sequences tending to zero; p. 22) is separable. As a countable dense set we may choose the set consisting of all sequences of rational numbers in which only a finite number of elements are different from 0 (e.g., $\{1, 0, 0, \dots\}$ or $\{\frac{2}{3}, -\frac{5}{2}, \frac{3}{4}, 0, 0, \dots\}$). This set is countable because the set of all finite subsets of the rational numbers is countable (p. 12). To show that it is dense in c_0, we recall that the distance between two points $x = (x_1, x_2, \dots)$ and $r = (r_1, r_2, \dots)$ in c_0 is $\sup|x_k - r_k|$. Let x be any point in c_0. We want to select a point r, where all r_k are rational and only finitely many of them are not zero, so that $\sup|x_k - r_k|$ is small. Taking an arbitrary positive ϵ to measure the desired degree of smallness, choose N so large that $|x_n| < \epsilon$ for $n > N$ (cf. the more detailed discussion of convergence in §8). Let $r_1, r_2, \ldots, r_N$ be rational numbers such that $|x_n - r_n| < \epsilon$ for $n = 1, 2, \ldots, N$. Then $r = (r_1, r_2, \ldots, r_N, 0, 0, \ldots)$ is the required point.

In the space C (continuous functions) we shall show later (§19) that the set of all polynomials is everywhere dense.

Exercise 6.6. Show that there are uncountably many polynomials.

Exercise 6.7. Show that there are countably many polynomials all of whose coefficients are rational.

Exercise 6.8. Show that if the set of all polynomials is dense in C, so is the set of all polynomials all of whose coefficients are rational. Deduce that C is separable.

An example of a space that is not separable is the space m of bounded sequences of real numbers (p. 23). We can see this as follows.

Exercise 6.9. Show that there is an uncountable set S in m, all the points of S being represented by sequences containing only 0's and 1's.

Exercise 6.10. What is the distance in m between any two of the points of the set S of Exercise 6.9?

Take any everywhere dense set E in m. We shall put the set S (described in Exercise 6.9) into one-to-one correspondence with a subset of E. This will show that E contains an uncountable subset and so is uncountable. Therefore m can contain no countable dense set. To put S into one-to-one correspondence with a subset of E, we proceed as follows. Take any point p of S. There is a point q of E at distance less than $\frac{1}{2}$ from p, since E is everywhere dense. The distance from q to any other point s of S is more than $\frac{1}{2}$, since $1 = d(p,s) \leqslant d(p,q) + d(q,s) < \frac{1}{2} + d(q,s)$. In this way we have associated a different point of E with each point of S, so that E has at least as many points as S, and hence uncountably many.

Exercise 6.11. Show similarly that the space B of §4 is not separable.

7. Compactness. It is frequently desirable to be able to assert that a set possesses a limit point, even though we may not be able to say just where that limit point is. Suppose, for example, that we are trying to prove that a bounded continuous real-valued function f, defined on a given set E in R_1, has a maximum. That is, we want to show that there is a point x of E such that $f(x)$ is actually equal to the least upper bound of all numbers $f(y)$ for y in E. (The reader is supposed to have at least a rough idea already of what a continuous function is; the formal definition will be given and discussed in §13.) We are then trying to show that there is an x in E such that $f(x) \geqslant f(y)$ for all y in E.

We supposed that f is bounded, that is, that the values $f(y)$ for y in E form a bounded set of real numbers. Such a set has a finite least upper bound M, so $f(y) \leqslant M$ for all y in E. There must then be points x_n in E such that $f(x_n) > M - 1/n$, since otherwise some smaller number than M would also be an upper bound. Moreover, we can suppose for the same reason that the x_n are all different, if E is not a finite set (and if E is finite there is nothing to prove). If the x_n have a limit point x in E, the continuity of f makes $f(x) \geqslant M$ because $f(x_n) \geqslant M - 1/n$. Since $f(x) \leqslant M$ in any case, by the definition of M, it follows that $f(x) = M$.

Now under what circumstances can we assert that a set $\{x_n\}$ of infinitely many different points of E has a limit point in E? This cannot always be true: for example, in R_1 the set $E_1 = \{\frac{1}{2}, \frac{1}{4}, \frac{1}{8}, \frac{1}{16}, \dots\}$ has the limit point 0, but this limit point is not in E_1. The set $E_2 = \{1, 2, 3, 4, \dots\}$ has no limit point in R_1. The *Bolzano-Weierstrass theorem* exists for the purpose of furnishing simple conditions that ensure that a set contains a limit point. It says that *an infinite bounded set in R_1 has a limit point in R_1*; if the set is also closed, it then contains a limit point. Assuming the truth of this theorem, for the moment, we see that *a bounded continuous real-valued function attains a maximum on any closed bounded set in R_1*. The examples (i) $f(x) = x$, with $E = R_1$; (ii) $f(x) = 1/x$, with E the interval $(0, \infty)$, show that both "closed" and "bounded" are essential conditions on the set E. (We shall see immediately that the boundedness of the *function* is redundant; it follows from the other hypotheses.)

Exercise 7.1. Deduce the preceding remark from the Bolzano-Weierstrass theorem.

We may prove the Bolzano-Weierstrass theorem by a process that has been suggested as a method for catching a lion in the Sahara Desert. We surround the desert by a fence, and then bisect the desert by a fence running (say) north and south. The lion is in one half or the other; bisect this half by a fence running east and west. The lion is now in one of two quarters; bisect this by a fence; and so on: the lion ultimately becomes enclosed by an arbitrarily small enclosure. The idea is actually employed in the Heligoland bird trap.[2b]

The essential point in applying this idea to our problem is that if a set E has an infinite number of points and lies in some finite interval I, then at least one half of I must contain an infinite number of points of E. Let I_2 be one of the halves, containing an infinite number of points of E, and bisect I_2. Again, one of the halves of I_2 contains an infinite number of points of E; call such a half I_3. Continue this process. We obtain a nested sequence of intervals $I_2, I_3, \ldots,$ each containing an infinite number of points of E. The left-hand endpoints of the intervals I_n form a set which is bounded above (since it is in I) and so has a least upper bound x. Every neighborhood of x contains some I_n, since the length of I_n tends to 0, and so contains an infinite number of points of E. That is, x is a limit point of E.

Exercise 7.2.   Prove the Bolzano-Weierstrass theorem for R_2.

As an application of the Bolzano-Weierstrass theorem together with the uncountability of the set of real numbers, we prove a theorem about the approximation of a function by the partial sums of its power series. *Let $f(x) = \sum_0^\infty a_k x^k$, with the series converging in $|x| < 1$. Suppose that, for each x in $[0, 1)$, $f(x)$ coincides with some partial sum of its power series, that is, for each x there is an n such that $\sum_{k=n+1}^\infty a_k x^k = 0$. Then f is a polynomial.*[3]

Let E_n be the set of points x in $[0, \frac{1}{2}]$ for which $\sum_{k=0}^n a_k x^k = f(x)$. Since there are countably many integers and uncountably many points in $[0, \frac{1}{2}]$, some E_n must be uncountable, and so infinite. Then E_n has a limit point in $[0, \frac{1}{2}]$, and f coincides on E_n with a polynomial, the same at every point of E_n. But an analytic function cannot coincide with a polynomial on a set having a limit point inside the interval of convergence without being itself a polynomial.

The statement that every infinite bounded set has a limit point makes sense in any metric space, although it may fail to be true. For example, it fails for the space consisting of the rational points of R_1 with the R_1 metric. We can see this by considering the set consisting of the rational approximations 1, 1.4, 1.41, 1.414, 1.4142, . . . to $\sqrt{2}$. This set is bounded; it is closed (since $\sqrt{2}$ is not in the space); but it contains no limit point. The Bolzano-Weierstrass theorem fails here because the space has, so to speak, too few points. It can also fail for a space that has too many points. For example, take the space B whose points are bounded functions on $[0, 1]$. We have seen (p. 43) how to construct an infinite number of such functions, each at distance 1 from the others; such a set of points of B cannot have a limit point.

A set E such that every infinite subset of E has a limit point in E was formerly called compact; we have just seen that bounded closed sets in R_1 or R_2 have this property. However, the term *compact* is now usually applied to sets with a less intuitive property (formerly called bicompactness). A set is called compact if whenever it is covered by a (not empty) collection of open sets, it is also covered by a finite subcollection of these sets (or if it is empty). (To say that E is covered by a collection of sets $\{G\}$ means that each point of E is in at least one G.)

To see how the property of compactness can be used, we shall use it to show that *a continuous real-valued function f defined on a compact set E in a metric space has a maximum on E*. First we show that the function is bounded. To each x in E we assign a neighborhood N with center at x such that $f(y) < f(x) + 1$ for all y in N. We can do this because f, being continuous, does not change much if we change x only a little. These neighborhoods are open sets, and every x is in at least one of them, so since E is compact a finite number of them cover E. Let these be $N_1, N_2, \ldots, N_n$. If x_k is the center of N_k, $f(x)$ cannot exceed the largest of the *finite* class of numbers $f(x_k) + 1$; so f is bounded above. Similarly f is bounded below.

Now we proceed to suppose that f does not attain a maximum on E, and deduce a contradiction. The values $f(x)$ for x in E form a bounded set, as we have just seen, so they have a least upper bound M, which we are supposing is not attained. To each x we can then assign a neighborhood N such that $f(y) < f(x) + \frac{1}{2}(M - f(x))$ for all y in N (by the continuity of the function). Again a finite number of these neighborhoods, $N_1, N_2, \ldots, N_n$ (not the same neighborhoods as before) cover E. Let $M'(< M)$ be the largest value of $f(x_k)$ where x_k is the center of N_k. Then for every y in E we have, taking the x_k that is in the same N_k as y,

$$f(y) \leqslant f(x_k) + \tfrac{1}{2}(M - f(x_k))$$
$$= \tfrac{1}{2}f(x_k) + \tfrac{1}{2}M < \tfrac{1}{2}(M' + M).$$

Thus the values of $f(y)$, for y in E, have the upper bound $\frac{1}{2}(M' + M)$, which is less than M, contradicting the definition of M as the least upper bound of f.

Exercise 7.3. If E is a set in R_1 and E is covered by a finite number of open intervals, we can reduce the number of intervals so that no point of E is in more than two of them and the reduced set still covers E.

Exercise 7.4. Show that a closed subset of a compact set is compact.

The preceding proof indicates that it is desirable to be able to recognize a compact set when we meet one. In R_1 this is easy: the *Heine-Borel theorem* states that *a set in R_1 is compact if it is closed and bounded*. The proof is almost the same as the proof of the Bolzano-Weierstrass theorem. Suppose that the conclusion of the Heine-Borel theorem is false. Then we have a set E which is closed and bounded, and a collection $\{G\}$ of open sets that covers E, whereas no finite subcollection of sets G covers E. The set E lies in some finite interval I; bisect I. The part of E in one of the halves of I must fail to be covered by a finite subcollection of the sets G; for, if both parts of E could be so covered, so could the whole set E. Let this half of I be I_2. Now bisect I_2 and continue the process as before. As with the Bolzano-Weierstrass theorem we see that every neighborhood of x contains an interval I_n that in turn contains a part of E that cannot be covered by a finite subcollection of the sets G (and hence is infinite). On the other hand, x is in E since E is closed, and so x can be covered by one of the sets G. Since G is open it contains a neighborhood of x, and this neighborhood contains an interval I_n if n is large enough. The part of E in this I_n is covered by a finite number (namely, one) of the sets G. We have thus arrived at a contradiction by supposing the Heine-Borel theorem false.

Exercise 7.5. Prove the Heine-Borel theorem in R_2.

Exercise 7.5a. Let S be a compact set in R_2, containing at least 3 noncollinear points. (a) Show that there is a triangle of largest area with vertices in S. (b) Does the diameter of S have to equal the length of one of the sides of this triangle?

The reader should have noticed that the hypotheses on the set in the Heine-Borel and Bolzano-Weierstrass theorems are the same. The similarity of both hypotheses and proofs should suggest a close relationship between the theorems. As a matter of fact, the theorems stand or fall together: for a given metric space, if either of them is true, so is the other. However, we shall omit the proof.[4]

Exercise 7.6. Show directly that the Heine-Borel theorem fails for the two examples given on p. 44 where the Bolzano-Weierstrass theorem fails.

Exercise 7.7. In R_1, let E be the interval $(0, 1]$. With each x associate the open interval $(\frac{1}{2}x, 2x)$. These intervals cover E. Show that no finite number can cover E, and explain why this fact does not contradict the Heine-Borel theorem.

Exercise 7.8. The set $[0, \infty)$ in R_1 is covered by the open intervals $(n - 1, n + 1)$, $n = 0, 1, 2, \ldots$. No finite number of these intervals cover the set. Explain why this fact does not contradict the Heine-Borel theorem.

Exercise 7.9. The set E in R_1 consisting of the rational numbers between 0 and 1 is not closed. It is covered by open sets in the following way: let x be covered by an open interval of length $\frac{1}{10}$ centered at x. A finite number of these open intervals do cover E. Explain why this fact does not contradict the Heine-Borel theorem.

Exercise 7.10. Let the set E in R_1 be the closed interval $[0, 1]$. Let each $x \neq 0$ in E be covered by the interval $[\frac{1}{2}x, 2x]$, and let 0 be covered by $[0, 0.1]$. The covering intervals are not open, yet a finite number of them do cover E. Explain why this fact does not contradict the Heine-Borel theorem.

It is also worth observing that not only is a set in R_1 (or in any R_k) compact if it is closed and bounded, but *if it is compact it is necessarily closed and bounded.* To see this, suppose that E is a nonempty compact set in R_1. It cannot be all of R_1, so its complement contains a point x. Consider all open finite intervals G such that the closure of G does not contain x. Among these intervals G are certainly neighborhoods of each point of E, since if $y \in E$, a neighborhood of y that reaches only halfway to x will not contain x in its closure. Therefore E is covered by the open sets G. Since E is compact, a finite collection of sets G covers E; let these sets be $G_1, \ldots, G_n$. Hence E is, in the first place, bounded, since it is contained in the union of a finite number of finite intervals. Since the closures of $G_1, \ldots, G_n$ do not contain x, the complements of these closures all do contain x, and therefore so does their intersection. The closures of the G_k are closed, their complements are open, so the intersection of the complements is open. Accordingly, x is in an open subset of $C(E)$; since x can be any point of $C(E)$, it follows that every point of $C(E)$ is in an open subset of $C(E)$. Therefore $C(E)$ is open (Exercise 5.15). Hence E is closed (Exercise 5.18).

8. Convergence and completeness. A great deal of analysis is concerned with infinite series or sequences of functions. The idea of an infinite series of numbers is more intuitive: we write, for instance, $\frac{1}{2} + \frac{1}{4} + \frac{1}{8} + \cdots$, meaning that we add up the terms one by one, thus forming the successive "partial sums"

$$\frac{1}{2}, \frac{1}{2} + \frac{1}{4}, \frac{1}{2} + \frac{1}{4} + \frac{1}{8}, \ldots,$$

and call the limit of these (if there is one) the sum of the infinite series. (It is assumed that the reader already has some idea of the meaning of limit; see, however, pp. 52–53.) In this case the partial sums are $\frac{1}{2}, \frac{3}{4}, \frac{7}{8}, \ldots$, and it is clear that the sum of the series must be 1. It is even clearer if we use the formula for the sum of a geometric progression:

$$\frac{1}{2} + \frac{1}{4} + \cdots + \frac{1}{2^n} = \frac{\frac{1}{2} - \frac{1}{2^{n+1}}}{1 - \frac{1}{2}} = 1 - 2^{-n}.$$

In general, if we write the infinite series

$$a_1 + a_2 + a_3 + \cdots,$$

we expect to calculate first a_1, then $a_1 + a_2$, then $a_1 + a_2 + a_3$, and so on, and call the limit (if any) of these partial sums the sum of the series. Note that, for example, $1 - 1 + 1 - 1 + \cdots$ and $(1 - 1) + (1 - 1) + (1 - 1) + \cdots$ must be different infinite series, since the first has successive partial sums $1, 0, 1, 0, \ldots$, whereas all the partial sums of the second are 0.

However, we have not actually defined "an infinite series" or even a particular infinite series: we have merely suggested a way of attaching a numerical value to a formula that has no meaning a priori. In order to see how we could actually give a definition, we notice that what was really used in the suggested calculation is the *sequence* of partial sums. Technically speaking a sequence is a function from the positive integers to some space: see §12. However, in less formal language we may think of a sequence of numbers as being a collection of numbers that have been labeled with the positive integers, preserving their order, and are not necessarily all different; thus $1, 0, 1, 0, \ldots$ is a sequence (the labeling being implicit); generally a sequence can be written in some such way as $a_1, a_2, a_3, \ldots$, or $\{a_n\}_1^\infty$, or simply $\{a_n\}$ if it is clear from the context that we are talking about a sequence and not a set. We must always distinguish a sequence from the set of numbers that appear in it. Any countably infinite set can be arranged as a sequence (in many ways), but a sequence need have only a finite number of different elements in it. (Note that a "finite sequence" such as $\{5, 12, 13\}$ is not a sequence according to our definition.) We can now define the infinite series $a_1 + a_2 + a_3 + \cdots$ as meaning no more and no less than the sequence whose elements are the partial sums

$$s_1 = a_1,$$
$$s_2 = a_1 + a_2,$$
$$s_3 = a_1 + a_2 + a_3,$$

and so on.

Conversely, any sequence of numbers defines a corresponding infinite series of which it is the sequence of partial sums. For example, the sequence $1, 0, 1, 0, \ldots$ is the sequence of partial sums of the series $1 - 1 + 1 - 1 + \cdots$. Generally, the sequence $s_1, s_2, s_3, \ldots$ is the sequence of partial sums of the series $s_1 + (s_2 - s_1) + (s_3 - s_2) + \cdots$.

The notion of sequence is more general than the notion of series since we can have a sequence whose elements are sets, or, indeed, points of any space we like; but there is no associated series unless there is an operation of addition for points of the space.

We shall naturally say that *an infinite series converges if its associated sequence of partial sums converges*; *otherwise the series is said to diverge*. To make this definition precise we must therefore define what we are to mean by the convergence of a sequence. If $\{s_n\}$ is a sequence of real numbers, we say that it converges to the limit L if $|s_n - L|$ eventually becomes and remains as small as we please. We then write $s_n \to L$. In more formal language, $s_n \to L$ if, given any positive number ϵ (with emphasis on its possible smallness), there is an integer N (usually rather large) such that $|s_n - L| < \epsilon$ provided that $n > N$. This definition extends immediately to sequences whose elements are points of any metric space: we have only to replace $|s_n - L|$ by $d(s_n, L)$. Thus the sequence of points $\{(\cos(1/n), \sin(1/n))\}$ of R_2 converges to $(1, 0)$; if elements x_n of the space C are defined by $x_n(t) = t^n(1 - t)^n$, $0 \leqslant t \leqslant 1$, the sequence $\{x_n\}$ converges to the element 0 of C (since $t(1 - t) \leqslant \frac{1}{4}$).

Although defining the infinite series $a_1 + a_2 + \cdots$ to mean the sequence $\{s_1, s_2, s_3, \ldots\}$ of its partial sums seems natural enough, we are under no compulsion to use this definition, and indeed it is not always the most reasonable definition to use. For some purposes it is better to define $a_1 + a_2 + \cdots$ to mean some other sequence, for example,

$$\frac{s_1}{1}, \frac{s_1 + s_2}{2}, \frac{s_1 + s_2 + s_3}{3}, \ldots.$$

or

$$\frac{s_1 + s_2}{2}, \frac{s_2 + s_3}{2}, \frac{s_3 + s_4}{2}, \ldots.$$

It can be shown that either of these definitions preserves the sum of any convergent series.[5] In addition, either one makes some divergent series converge; for example, the divergent series $1 - 1 + 1 - 1 + \cdots$ has $s_1 = 1$, $s_2 = 0$, $s_3 = 1$, and so on, so that either of the suggested definitions would give it the sum $\frac{1}{2}$.

We are now going to discuss some properties of sequences of points in a metric space; the idea of infinite series has served to motivate the introduction of sequences, but we have no further use for infinite series just now. (Some theorems are more conveniently formulated in terms of series than in terms of sequences; we shall have an example in §22.)

If a sequence converges to a limit L, its elements eventually become and remain close to each other. Indeed, let N be so large that, for $n > N$, we have $d(s_n, L) < \epsilon/2$; let $m > N$; then $d(s_m, L) \leqslant \epsilon/2$ also. By the triangle inequality, $d(s_m, s_n) \leqslant d(s_n, L) + d(s_m, L) < \epsilon$. In other words, $d(s_m, s_n)$ can be made as small as we please by taking m and n simultaneously sufficiently large.

If a sequence $\{s_n\}$ has the property that its elements eventually become and remain close to each other, in the sense just described, it is called a *Cauchy sequence*, and is said to *converge*. It may or may not converge to a limit in the space. For example, in the metric space of rational numbers with the R_1 distance, the sequence

$$\{1, 1.4, 1.41, 1.414, 1.4142, \ldots\}$$

of decimal approximations to $\sqrt{2}$ converges. In fact, if $m > N$ and $n > N$, s_m and s_n agree at least through the Nth decimal place and so $|s_m - s_n| < 10^{-N}$. However, the sequence does not converge to a point of the space.

A metric space for which every Cauchy sequence converges to a point of the space is called *complete*. The metric space of rational numbers is not complete. However, R_1 is complete, as we shall see shortly. It is, in fact, always possible to make a metric space complete by adding new points to it to make a larger space, somewhat as the real numbers can be constructed from the rational numbers. We shall not discuss this construction here, however.[6]

We shall now show that the completeness of R_1 follows from the least upper bound property that we took as fundamental in §2. Let $\{s_n\}$ be a Cauchy sequence. Then if ϵ is a given positive number there is an N such that $|s_m - s_n| < \epsilon$ if $n > N$ and $m > N$. In the first place, the different numbers s_n must form a bounded set. To see this, take $\epsilon = 1$, find the corresponding N, and take a convenient $m > N$. Then $s_n = s_m + (s_n - s_m)$, whence $|s_n| \leqslant |s_m| + 1$ if $n > N$. Since the finite set

$$\{s_1, s_2, s_3, \ldots, s_N\}$$

is bounded, so is the whole set of numbers s_n.

Now let L_k be the least upper bound of the set consisting of all the different s_n for $n > k$. Since taking a larger k means that we are considering the least upper bound of a smaller set of numbers, we have $L_k \geqslant L_{k+1} \geqslant \cdots$. Let L be the greatest lower bound of the set of all L_k. We shall show that $s_n \to L$. Let ϵ be any positive number, and take the N corresponding to this ϵ, so that if $n > N$ we have $|s_m - s_n| < \epsilon$. By the definition of greatest lower bound there must be an L_k, with $k > N$, between L and $L + \epsilon$. Since L_k is the least upper bound of the set of s_n with $n > k$, there is an s_m (with $m > k$) between $L_k - \epsilon$ and L_k; hence $L - \epsilon \leqslant s_m \leqslant L + \epsilon$. Since s_n differs from s_m by at most ϵ, we have $L - 2\epsilon \leqslant s_n \leqslant L + 2\epsilon$. Since 2ϵ is just as arbitrary as ϵ, we now know that s_n can be made arbitrarily close to L by taking n large enough; that is, $s_n \to L$.

Exercise 8.1. Show conversely that if the completeness of R_1 is assumed, the least upper bound property follows.

Exercise 8.2. If $\{s_n\}$ is a sequence of points of R_1 such that $s_n \leqslant s_{n+1}$, and $s_n \leqslant M$ for all n, then $\{s_n\}$ converges. In other words, *an increasing bounded sequence has a limit*. (This is often taken as the basic form for the completeness of R_1.)

Exercise 8.3. Prove that R_2 is complete.

We shall see in §17 that the space C is complete; and, more generally, the space of continuous functions on any given compact set is complete.

We noticed (Exercise 4.2) that a subset of a metric space is again a metric space (if the same metric is used).

Exercise 8.4. A subset of a complete metric space is not necessarily a complete metric space. Give an example of this.

However, we can show that *a nonempty closed subset E of a complete metric space S is also a complete metric space* (again, using the original distance). Let $\{x_k\}$ be a Cauchy sequence of points of E. It is also a Cauchy sequence of points of S, since the distances in E and S are the same. Since S is complete, $x_k \to x_0$, where $x_0 \in S$. We have only to show that $x_0 \in E$. There are two cases to consider. In the first case, all but a finite number of the x_k coincide. Clearly they must then coincide with x_0, which accordingly belongs to E. In the second case there are an infinite number of different x_k's. The condition $x_k \to x_0$ then shows that x_0 is a limit point (in S) of the set consisting of these x_k, and hence x_0 is a limit point of the set E (considered as a subset of S). Since E is closed, it contains its limit points; so $x_0 \in E$.

We have to distinguish carefully between the limit of a sequence and a limit point of the set consisting of the (different) elements of the sequence. For example, the sequence $\{0, 0, 0, \ldots\}$ in R_1 has the limit 0, but the set of elements of the sequence has only one point and so has no limit point. On the other hand, the set of elements of the sequence $\{0, 1, \frac{1}{2}, \frac{1}{2}, \frac{1}{4}, \frac{3}{4}, \frac{1}{8}, \frac{7}{8}, \ldots\}$ in R_1 has two limit points (0 and 1), but the sequence has no limit. The necessity of this distinction was what forced us to consider two cases in the preceding proof.

However, there is a close connection between the limit of a sequence and the limit points of the set consisting of the different elements of the sequence.

Exercise 8.5. Show that if a sequence converges to L and has an infinite number of different elements, the set consisting of all the different elements of the sequence has L as a limit point, indeed as its only limit point.

Exercise 8.6. Hence show that if a sequence converges to a limit and its elements belong to a closed set, the limit of the sequence belongs to the same set.

Exercise 8.7. Hence show that if E is a nonempty compact set in R_1, E has a largest element.

We may conclude from Exercise 8.5 that if a sequence converges, the set of its elements does its best to have the limit of the sequence as a limit point. In the other direction, we can easily see that *if a set E has a limit point L, there is a sequence of points of E having L as its limit*. In fact, there is a point x_1 in E and distant less than 1 from L; then there is a point x_2 distant less than $\frac{1}{2}$ from L; and so on. In applications the set E often consists of the elements of a sequence; if these elements have a limit point, there is a subsequence converging to the limit point. We shall refer to this as the *subsequence principle*; it has many uses.

For example, one of the ways to prove the fundamental theorem of algebra is to show (A) the absolute value of a nonconstant polynomial $P(z)$ has no minimum in the complex plane, so that there is a sequence $\{z_n\}$ on which $P(z_n) \to 0$; (B) by the subsequence principle, a subsequence of $\{z_n\}$ has a limit z_0 and hence $P(z_0) = 0$. Some nineteenth century proofs stop after step (A), apparently regarding (B) as self-evident.

Exercise 8.8. If E is any bounded sequence in R_2 ("bounded" means that the elements of E form a bounded set), show that E contains at least one convergent subsequence. (This is an analogue for sequences of the Bolzano-Weierstrass theorem for sets.)

As an illustration of the use of the subsequence principle, we discuss the diameter of a set E. The *diameter* is defined to be the least upper bound of the distances between points of E; in symbols, diam $E = \sup d(x, y)$ for x and y in E. For example, in R_2 the diameter of a circle of radius 1 is 2; so is the diameter of the (open) area inside the circle and so is the diameter of the closed area. The diameter of the set consisting of the three points $(0,0)$, $(0,1)$, and $(1,0)$ is $\sqrt{2}$. It may happen that there are no points (x, y) in E for which $d(x, y) = $ diam E, even when E is bounded and so has a finite diameter. For example, when E is a neighborhood in R_2 there are no two points of E that are diam E apart.

However, there certainly are such points if E is a compact nonempty set in R_1 or R_2. We give a proof by contradiction. Suppose that we have not succeeded in finding points x and y of E for which $d(x, y) = $ diam E. Then we must be able, by the definition of diameter, to find pairs of points $x_1, y_1; x_2, y_2; \ldots$ such that $d(x_n, y_n) > $ diam $E - 1/n$. An infinite number of the x_n or of the y_n will be different (otherwise we have already found x and y such that $d(x, y) = $ diam E). If there are an infinite number of different x_n they have a limit point (by the Bolzano-Weierstrass theorem), and we can select a subsequence having this limit point as a limit. If there are only a finite number of different x_n, one of them, say x_1, will occur an infinite number of times, and then a sequence all of whose elements are equal to x_1 will have x_1 as a limit. We can proceed similarly with the y_n that correspond to the x_n already selected. The result is that we have sequences, which for simplicity of notation we may call $\{x_n\}$ and $\{y_n\}$ again, such that $x_n \to x_0$, and $y_n \to y_0$, and $d(x_n, y_n) \to \operatorname{diam} E$. Then we must have $d(x_0, y_0) = \operatorname{diam} E$. For, on the one hand $d(x_0, y_0)$ cannot exceed $\operatorname{diam} E$, since E is closed and so x_0 and y_0 are in E. On the other hand, the triangle inequality shows that

$$d(x_n, y_n) \leqslant d(x_n, x_0) + d(y_n, y_0) + d(x_0, y_0),$$

so that $\operatorname{diam} E \leqslant d(x_0, y_0)$.

Exercise 8.9. Define the distance between two sets F and G to be $\inf d(x, y)$ for x in F and y in G. Show that if F and G are in R_2, if F and G are closed and not empty, and F is bounded, then there are points x in F and y in G such that $d(x, y)$ is the distance between F and G.

Exercise 8.10. If N is a neighborhood of y, consisting of all x such that $d(x, y) < r$, is $\operatorname{diam} N = 2r$? (Consider (a) R_1 or R_2; (b) general metric spaces.)

Exercise 8.11. Show that E and its closure have the same diameter.

9. Nested sets and Baire's theorem. Suppose that we have two sets E_1 and E_2, that $E_1 \supset E_2$, and that E_2 is not empty. Then there is at least one point that is simultaneously in both sets, since $E_1 \cap E_2 = E_2$. Similarly, if we have a finite number of sets that are *nested*: $E_1 \supset E_2 \supset E_3 \supset \cdots \supset E_n$, and if the last set E_n is not empty, there is at least one point that is simultaneously in all the sets. There is nothing corresponding to this when we have an infinite number of nested sets, none of which is empty; the intersection of all the sets may quite well be empty nevertheless. Consider the following three examples: (i) E_n is the open interval $(0, 1/n)$ in R_1; (ii) E_n is the set of those points x in the metric space of rational points in R_1 that are subject to the inequality $|x - \sqrt{2}| < 1/n$; (iii) E_n is the interval $[n, \infty)$ in R_1. In each of these cases the intersection of all the sets E_n is empty.

We now give conditions that prevent a nested collection of sets from having an empty intersection. *Cantor's nested set theorem*: *If $E_1 \supset E_2 \supset E_3 \supset \cdots$; if the E_n are closed and not empty; if the underlying space is complete; and if* diam $E_n \to 0$; *then there is exactly one point in the intersection of all the E_n.*

Cantor's theorem involves three conditions besides the condition that the sets are nested and not empty: closure and small diameter of the sets, and completeness of the space.

In each of our three examples of nested sets with empty intersection, a different one of these conditions fails.

To prove Cantor's theorem, let x_n be any point belonging to E_n. The sequence $\{x_n\}$ is a Cauchy sequence since if $m > n$, $x_m \in E_n$ and $d(x_n, x_m) \leqslant$ diam E_n, which approaches zero. Since the space is complete, $\{x_n\}$ has a limit in the space. If we select any E_n, all the x_k belong to E_n if $k \geqslant n$, so the limit belongs to E_n because E_n is closed. That is, the limit is in every E_n. Finally, there cannot be two points that are in every E_n, since diam E_n is at least as large as the distance between any two of the points of E_n.

It is sometimes useful to have the following weaker theorem: *if we keep all the hypotheses of Cantor's theorem except that we no longer require that* diam $E_n \to 0$, *but require instead that the E_n are compact, we still can say that the intersection of the E_n is not empty* (although it may now contain more than one point). Since we have kept the hypothesis that E_n is closed, in any R_k our new hypothesis amounts to supposing that E_n is also bounded. In R_k, the generalized theorem is an easy application of the subsequence principle: a sequence $\{x_n\}$ consisting of one point from each set has a subsequence that has a limit, and this limit is a point of the required kind.

In the general case we have to proceed differently. Let us cover E_1 by neighborhoods of all its points, each of diameter at most 1. Because E_1 is compact, a finite number of these, say $N_1, \ldots, N_p$, cover E_1. One of the N_k must contain points of all the E_n (from $n = 1$ onward). Otherwise each N_k would miss some E_m, and hence miss all E_n with $n > m$ (because the E_n are nested). If N_1 is disjoint from E_{m_1}, and N_2 is disjoint from E_{m_2}, and so on, no N_k contains points of E_n for $n > m_0$, the largest of $m_1, m_2, \ldots, m_p$. Since the N's cover E_1, this would mean that E_1 contains no points of E_n for $n > m_0$, contradicting the assumption that the E_n are nested. Thus it must be true, as asserted, that some N contains points of all the E_n for $n = 1, 2$, and so on. Then the closures of the sets $N \cap E_n$ are closed, nested, and of diameter at most 1. Repeat this reasoning with neighborhoods of diameter at most $\frac{1}{2}$, covering the closure of $N \cap E_1$; then with neighborhoods of diameter at most $\frac{1}{3}$; and so on. We obtain nested subsets of the E_n with diameters approaching zero, and we can then apply Cantor's theorem in its original form.

We can sometimes use Cantor's theorem to show that a set in which we are interested cannot be empty. If we want to know that there are things with a certain property, we can be sure that there are some if we can exhibit them as the intersection of a nested collection of sets satisfying the hypotheses of Cantor's theorem. However, it is often more efficient not to use the nested sets directly, but to use instead another theorem that is a consequence of Cantor's theorem. To state this new theorem we need to introduce a new class of sets, namely, sets that can be represented as unions of countably many sets, each nowhere dense. ("Countably many" includes none, or one, or a finite number, as well as a countable infinity.) A set that can be represented as the union of countably many nowhere dense sets is called a set of *first category*. (Since the name is not at all descriptive, the name *meager set* has been suggested as an alternative; the reason for using this name will appear shortly.)

In R_1, any set consisting of a finite number of points is of first category. So is any countable set, for example, the set of all rational numbers, since although this set is everywhere dense, it is the union of a countable number of sets, each consisting of a single point. The Cantor set, being nowhere dense, is of first category but uncountable. If we form the union of the Cantor set with the set of all rational points, we obtain a set of first category which is both everywhere dense and uncountable.

Sets that are not of first category are said to be of *second category*. Since the empty set is nowhere dense, it is of first category; hence *a set of second category cannot be empty*. This fact is the basis for the principal use of the notion of category: if we can show that a set is of second category it must contain points. We can sometimes exhibit the aggregate of things of a particular kind as a set of second category; there must then be things of this kind. Some examples will be given presently. The technique for applying this idea depends on *Baire's theorem*, which states that *a complete metric space is of second category*.

Before proving Baire's theorem we make a few remarks. First, the completeness of the metric space is an essential part of the theorem. The metric space whose points are the rational points of R_1, with the R_1 distance, is not complete; each point of the space, regarded as a set, is nowhere dense; the whole space is therefore the union of a countable number of nowhere dense sets.

We cannot state without reservation that a countable set is of first category, although the preceding example may make this seem plausible. Unfortunately, as we noted in Exercise 6.1, a single point need not form a nowhere dense set. This happens, in particular, in any space that contains only a finite number of points. The next exercise considers the other extreme.

Exercise 9.1. Show that if all the points of a space are limit points, any set containing only a single point is nowhere dense.

Exercise 9.2. Hence show that Baire's theorem implies that both R_1 and the Cantor set are uncountable.

We now prove Baire's theorem. Let $\{E_n\}$ be a sequence of nowhere dense sets in a complete metric space. We are to prove that there is at least one point of the space that is in none of the E_n. The idea of the proof is that since E_1 is nowhere dense, its complement contains a neighborhood N_1; N_1, in turn, contains a subneighborhood N_2 that is in the complement of E_2 as well as in the complement of E_1; and so on. In this way we obtain a nested sequence of neighborhoods that are disjoint from more and more of the E_k, and a common point of all of them cannot be in any E_k.

To show that there is actually a common point, we must take a certain amount of care so that we can apply Cantor's theorem. Select first a neighborhood N_1 in $C(E_1)$. Take a concentric subneighborhood of diameter at most 1, and let M_1 be the closure of this subneighborhood. Next, since E_2 is nowhere dense, M_1 contains a neighborhood that is in $C(E_2)$ (as well as in $C(E_1)$). Let M_2 be the closure of a concentric subneighborhood of N_2 whose diameter is less than $\frac{1}{2}$. Continuing in this way we obtain nested closed sets M_k, whose diameters tend to 0, which are not empty, and which have the property that M_k is disjoint from $E_1, E_2, \ldots, E_k$. The common point of all M_k is a point of the required kind, since it cannot be in any E_k.

Exercise 9.3. If E is a nonempty bounded subset of R_2, then E and the boundary of E have the same diameter.

10. Some applications of Baire's Theorem

(i) A PROPERTY OF REPEATED INTEGRALS. Let f be a continuous real-valued function on a real interval, say [0, 1]. Let f_1 be any integral of f, f_2 any integral of f_1, and so on. If some f_k vanishes identically, so does f: we have only to differentiate f_k repeatedly. The following proposition generalizes this simple fact: *if for each x there is an integer k, possibly differing from one x to another, such that $f_k(x) = 0$, then f vanishes identically.*

To prove this theorem, let E_k be the set of points x for which $f_k(x) = 0$; then our hypothesis says that every x in [0, 1] is in some E_k. By Baire's theorem, not every E_k is nowhere dense. Hence there is some k for which the closure of E_k fills an interval I_k. For this particular k, since f_k is continuous and vanishes on E_k, we must have $f_k(x) = 0$ for every x in I_k. If I_k is not all of [0, 1], we repeat this argument with any remaining part of [0, 1], and so on. In this way we have $f(x) = 0$ for all points x of an everywhere dense set; and since f is continuous, it then follows that $f(x) = 0$ for every x in [0, 1].

Thus if $f(x) \not\equiv 0$, then no matter how the integrals f_k are selected, there must be some x (indeed, an everywhere dense set) such that $f_k(x) \neq 0$ for every k.

(ii) A CHARACTERIZATION OF POLYNOMIALS. Consider again a continuous real-valued function f on [0, 1]. If f has an nth derivative that is identically zero, it is easily proved, for example, by repeated application of the law of the mean (and see p. 177) that f coincides on $[0, 1]$ with a polynomial (of degree at most $n - 1$). The following theorem generalizes this in the spirit of example (i). *Let f have derivatives of all orders on $[0, 1]$, and suppose that at each point some derivative of f is zero. That is, for each x there is an integer $n(x)$ such that $f^{(n(x))}(x) = 0$. Then f coincides on $[0, 1]$ with some polynomial.*[7]

We can start the proof just as in (i). Let E_n be the set of points x for which $f^{(n)}(x) = 0$. By hypothesis every x is in at least one E_n. By Baire's theorem there is a closed interval I in which some E_n is everywhere dense. Since $f^{(n)}$ is a continuous function, $f^{(n)}(x) \equiv 0$ in I and f coincides in I with a polynomial. If I is not all of $[0, 1]$, repeat the reasoning in any remaining part of $[0, 1]$, and so on. In this way we see that there is an everywhere dense set of intervals in each of which f coincides with a polynomial. We still have to show that f coincides with the same polynomial in all the intervals.

To do this, we are going to apply Baire's theorem again to the nowhere dense set H that is left when we remove the interiors of our dense set of intervals from $[0, 1]$. We first need to show that H is perfect. In the first place H is closed, since it is obtained by removing a collection of open intervals from a closed interval. Suppose that H is not perfect, and not just the pair $\{0, 1\}$ (otherwise there was only one interval to begin with and there is nothing more to prove). Then H must have a point y that is not a limit point. This point is the common endpoint of two intervals in each of which f coincides with some polynomial. Then if n exceeds the degree of both polynomials, $f^{(n)}(x) = 0$ for x in both intervals, and at the endpoint by the continuity of $f^{(n)}$. Therefore f coincides with a polynomial in the union of the two intervals, and the point y does not belong to H after all.

The preceding discussion shows that H is perfect, and we may again suppose it not to be empty. Consider H as a complete metric space. By Baire's theorem for H, some E_n is everywhere dense in some neighborhood in H, that is, in the part of H that is in some interval J. In other words, there is an interval J that contains points of H, and $f^{(n)}(x) = 0$ for every x in $J \cap H$ (with the same n). Now J also contains intervals complementary to H, and in each such interval K we have $f^{(m)}(x) = 0$ for some m (depending on K). If $m \leqslant n$, we have $f^{(n)}(x) = 0$ in K, by differentiating. If $m > n$, we have $f^{(n)}(x) = f^{(n+1)}(x) = \cdots = 0$ at the endpoints of K, since these are points of H. Hence, by integrating $f^{(m)}$ repeatedly, we get $f^{(n)}(x) = 0$ throughout K. This reasoning applies to every interval K that is complementary to H and is in J; so $f^{(n)}(x) = 0$ throughout J. Thus J contains no points of H after all. But we arrived at J as an interval containing points of H on the assumption that H was not empty. This contradiction means that H must be empty, there was only one interval $I = [0, 1]$ to begin with, and f coincides with a single polynomial throughout this interval.

(iii) CONTINUOUS EVERYWHERE OSCILLATING FUNCTIONS. In our next application of Baire's theorem the underlying metric space will be the space C of continuous functions on a real interval. It will be shown later (§17) that this space is complete. Let us seek first to construct a continuous function that is not monotonic in any interval. This can actually be done more directly, but it is a good illustration of the use of Baire's theorem in a fairly uncomplicated situation. The intervals with two rational endpoints form a countable set. Let them be $I_1, I_2,$

$I_3, \ldots,$ in some order, and let E_n be the set of elements of the space C that are monotonic on I_n. We are going to show that each set E_n is nowhere dense in C; it then will follow from Baire's theorem that there is an element of C that is not in any E_n. In other words, there is a continuous function that is not monotonic on any I_n, and hence not monotonic on any interval (since every interval in R_1 contains an interval with rational endpoints).

The technique for showing that E_n is nowhere dense is one that is useful in many applications: we show that $C(E_n)$ is open and everywhere dense.

Exercise 10.1. Show that a closed set with an everywhere dense complement is nowhere dense.

We first show that $C(E_n)$ is open. If f belongs to $C(E_n)$, f is not monotonic in I_n. This means that there are three points x, y, z in I_n, with $x < y < z$, and $f(x) < f(y)$ and $f(z) < f(y)$ (or else $f(x) > f(y)$ and $f(z) > f(y)$). Recalling that the distance between elements f and g of C is $\max|f(x) - g(x)|$, we see that if g is closer to f than half the smaller of the numbers $f(y) - f(x)$ and $f(y) - f(z)$, we also have $g(x) < g(y)$ and $g(z) < g(y)$, so that g is not monotonic in I_n. That is, all elements g sufficiently close to f are not in E_n, which is to say that $C(E_n)$ is open.

To say that the complement of E_n is everywhere dense is to say that in every neighborhood in C there exists a function f that is not monotonic in I_n. It is rather intuitive that there is a very wiggly function close to any given continuous function. To justify this intuition, we can (for example) let g be the center of the given neighborhood in C; we can approximate g in the metric of C, as closely as we like, by a polynomial p (§19). Now p has a bounded derivative, so if we add to p a small saw-tooth function with very steep teeth we obtain a function f that is as close as we like to g and is not monotonic in I_n.

(iv) EXISTENCE OF CONTINUOUS NOWHERE DIFFERENTIABLE FUNCTIONS. The fact that a continuous function may fail to have a derivative at any point came as a surprise to the mathematicians of the nineteenth century. However, it turns out that "most" continuous functions have this property, and we should rather be surprised that any continuous functions are differentiable at all. Still more surprising is the fact that a function may be everywhere oscillating and still have a finite derivative at every point. Unfortunately all known examples of this last phenomenon are too complicated to give here.[8]

What we are going to show is that[9] *the elements of the space C that have, even at one point, a finite derivative, even on one side* (cf. p. 131) *form a set of first category in C*. This theorem shows that all the functions that are ordinarily encountered in calculus form only a set of first category in C. We do not exclude the possibility that "most" elements of C might have infinite one-sided derivatives at most points (geometrically, their graphs might have cusps). We shall see later (§21) that a continuous function actually cannot have a vertical tangent at *all* the points of an interval. Although there really are nowhere differentiable functions that do not even have one-sided infinite derivatives, they are much harder to construct[10]; the greater difficulty may be connected with the fact (which we cannot prove here) that these functions form only a set of first category.[11] The "typical" continuous function has an everywhere dense set of cusps (like that of $y = |x|^{1/2}$ at the origin). A nowhere differentiable function must be everywhere oscillating, since a monotonic function has a derivative at most points (§22).

We now prove that nowhere differentiable functions form a set of second category in C. Consider the set E_n formed of all those elements f of C such that, for some point x in the interval $[0, 1 - 1/n]$

$$\left| \frac{f(x + h) - f(x)}{h} \right| \leqslant n$$

if $0 < h < 1/n$. Clearly a function f that has a finite right-hand derivative at x belongs to E_n for some n. Hence the union of all E_n contains all the elements of C that have a finite right-hand derivative at some point. We shall show that each E_n is nowhere dense; then the union of the E_n is of first category (and so cannot be all of C). As in the preceding example, we do this by showing that E_n is closed and has an everywhere dense complement.

That E_n is closed follows as in example (iii), or from the fact that the inequality defining E_n is preserved under convergence in C. That the complement of E_n is everywhere dense follows just as in (iii): if f is an arbitrary continuous function, we find a function close to f with a bounded derivative, and then add to this function a small continuous function the slope of whose graph has large absolute value.

(v) Decomposition of a Closed Interval

Exercise 10.2. Show that a closed interval cannot be the union of a countably infinite number of disjoint closed nonempty sets.[11a]

11. Sets of measure zero. A set of first category may be thought of as having relatively few points, principally because of the fact that it cannot exhaust a complete metric space. It may, on the other hand, seem rather large if looked at from other points of view. It may be everywhere dense, as the set of rational points in R_1 is. It may be uncountable, as the Cantor set is. It may be both everywhere dense and uncountable, as we observed on p. 62.

There is another—quite different—kind of "sparse" set that has many uses. Suppose that a subset E of R_1 has the property that it can be covered by a countable collection of open intervals whose total length is arbitrarily small. Then E is called a *set of measure zero*. There is a similar definition in any R_n. Sets of measure zero are just the sets that are negligible in the theory of Lebesgue integration, and the name comes from that theory. If something happens except on a set of measure zero, it is said to happen *almost everywhere*, or for almost all points.

The union of two, or of finitely many, or even of a countable infinity of sets of measure zero is again of measure zero.

Obviously a countable subset of R_1 is a set of measure zero, so a set of measure zero can be everywhere dense. However, a set of measure zero need not be countable, or even of first category; and, on the other hand, a set of first category may fail to be of measure zero. Let us justify these statements by examples. First, the Cantor set is uncountable but of measure zero. For, if ϵ is a small positive number, take enough complementary intervals of E so that their total length exceeds $1 - \epsilon/2$. The rest of the unit interval containing E can be covered by a finite set of intervals of length not exceeding ϵ and so E is certainly of measure zero.

To construct a set that is of first category, but not of measure zero, modify the construction of the Cantor set as follows. Let $\{a_n\}$ be a sequence of positive numbers such that $\sum a_n = \epsilon < 1$. Remove from the unit interval an open interval of length a_1; then from each remaining interval remove an open interval of length $\frac{1}{2} a_2$; and so on. As with the Cantor set, we see that the resulting set E is nowhere dense, hence of first category. On the other hand, E cannot be of measure zero, since if it were covered by a countable set of intervals of total length less than $1 - \epsilon$ we should have a unit interval covered by a set of intervals of total length less than 1.

To construct a set that is both of second category and of measure zero, construct a generalized Cantor set E of the kind just described. Then in the complementary intervals of E construct similar sets, and so on. We can arrange that the complement of the union of all the generalized Cantor sets has measure zero; but since each Cantor set is nowhere dense, this complement is of second category.

Since an interval in R_1 is not of measure zero, another way to show that a set of points is not empty is to exhibit it as the complement of a set of measure zero. For instance, any countable subset of R_1 is of measure zero, and so R_1 cannot be countable. A set of measure zero, like a set of first category, is "small" in the sense that its complement cannot be empty; to say anything more about how small it is requires the theory of Lebesgue measure.

Exercise 11.1. Let a subset E of R_1 have the following property: there is a positive q less than 1 such that, for every interval (a, b), the set $E \cap (a, b)$ can be covered by countably many intervals whose total length is at most $q(b - a)$. Then E is a set of measure zero. (Intuitively, this means that a set that covers at most a fixed fraction of every interval covers almost none of every interval.)

CHAPTER 2

FUNCTIONS

12. Functions. In elementary mathematics it is customary to say that y is a function of x if, when x is given, y is determined (uniquely; we are not concerned with "multiple-valued functions"). This is a good working definition and one that suffices for most practical purposes. However, we should realize that it does not define "function," although it does give a definite meaning to some phrases containing this word. (In a somewhat similar way, we are accustomed to attaching a definite meaning to the phrase "$y \to \infty$" even though ∞ by itself has no meaning.) However, it is interesting, and sometimes helpful, actually to define a function as a genuine mathematical entity. Consider two sets E and F of real numbers, neither of which is empty, and form a class of ordered pairs (x, y) with $x \in E$, $y \in F$, where each x occurs exactly once and each y occurs at least once. Such a class of ordered pairs is called a *function* with *domain* E and *range* F, or a function from E to F; or, on occasions when it is unnecessary to say precisely what F is, a function with domain E and values in R_1, or a function from E into R_1, or a real-valued function with domain E, etc. For example, let E be all of R_1, let F be the closed interval $[-1, 1]$, and let the ordered pairs be $(x, \sin x)$ for each x in R_1. This is what is usually spoken of as "the function $\sin x$." Notice that if we take E to be, for instance, the interval $[0, 2\pi]$ instead of all of R_1, the set of ordered pairs $(x, \sin x)$, $x \in E$, is a different function, the *restriction* of the original function to $[0, 2\pi]$.

For another example, let E consist of the positive integers $1, 2, 3, \ldots$, and let F be the set $\{1, 4, 9, 16, \ldots\}$. The function whose ordered pairs are $(1, 1)$, $(2, 4)$, $(3, 9), \ldots$ is an example of the special kind of function that we usually call a *sequence*, specifically a sequence of real numbers.

Exercise 12.1. An equation in x and y determines a set of ordered pairs (x, y) whose components satisfy the equation. Accordingly, it may (or may not) determine a function. For the following equations, decide which ones determine functions when all the ordered pairs of real numbers (x, y) satisfying the equations are taken.

    (a) $x^2 + y^2 = 25$,   (b) $x^2 + y^2 = 0$,
    (c) $y = |x|$,   (d) $|x| + |y| = -2$,
    (e) $y = \cos x$,   (f) $x = \cos y$.

If F consists of a single point, say the point 3, the function whose ordered pairs are $(x, 3)$ is a *constant* function; ideally, it should be distinguished notationally from the number 3.

The definition of function is easily extended to a more general setting. Let E be a nonempty set of points in some space S (which might be R_n, or C, or anything else; it need not be metric); let F be another nonempty set of points belonging to some space T, in general quite different from S. The class of all ordered pairs (x, y) with x in S and y in T is called the Cartesian product of S and T. A function from E to F is a subset of this Cartesian product consisting of all pairs (x, y) where each x in E occurs exactly once and each y in F occurs at least once. It is also called a function from E into T. We speak of the points of F as the *values* of the function, and of F as the *image* of E. It is often convenient to call the function a *mapping* of E on F.

The Cartesian product of two R_1's is the space R_2 (if we introduce the right metric in the product). A function whose domain and range are in R_1 is just the set of points of R_2 that are in what we ordinarily call the *graph* of the function. "A real function of a real variable" ordinarily means a function with domain and range in R_1. For the purposes of this book it is unnecessary to attempt to define a "variable," and we shall neither define nor use this word.

The advantage of the abstract definition of a function as a class of ordered pairs is that it gives us a mathematical object defined in terms of concepts we have already had. Every proof about functions can be phrased in terms of it. Its disadvantage is that it loses most of the intuitive content of the notion of function. For many purposes, and especially for a student's first introduction to the concept, it is better to think of a function as a mapping or transformation or operator (emphasizing the process by which values are obtained from points of the domain); as a rule (emphasizing the correspondence between the values and the points of the domain); or in physicists' language as a field (emphasizing the domain, and the association of a value with each point of the domain—usually in physics the domain is in R_3 and the range is in R_1 (scalar field) or in R_3 (vector field)). We should perhaps think of a class of ordered pairs as being a *model* of a function rather than the function itself. On the other hand, if "function" is taken as a primitive notion, then "ordered pair" can be defined, or modeled, in terms of it (as a function on the set $\{1, 2\}$).[11b]

A *sequence* is a function whose domain consists of all the positive integers. If we are talking about sequences, the

domain is understood to be fixed and so we can specify
the sequence by listing the points of the range in the order
imposed by the points of the domain. We then often speak
of the points of the range (repeated as often as necessary)
as the elements of the sequence, rather than as the values
of the function. This brings us back to the informal
definition of a sequence that we have already used (p. 51).
Thus "the sequence $\{2, 4, 8, 16, \ldots\}$" or "the sequence
$\{2^n\}$" means the set of ordered pairs $(1, 2), (2, 4), (3, 8)$,
$\ldots$. "The sequence $\{1\}$" means the set of ordered pairs
$(1, 1), (2, 1), (3, 1), \ldots$. A subsequence of a given sequence
is a restriction of the sequence to a subset of the integers;
it can be identified with the sequence obtained by
renumbering its elements. For instance, $\{2^n\}$ is a subsequence of $\{n\}$. (As noted on p. 52, a "finite sequence" is
not a sequence in the sense in which we use the word
sequence.)

In careful work it is usually helpful to distinguish
notationally between (say) f, the name of a function, and
$f(x)$, "the value of the function at the point x." In other
words, f is to denote a set of ordered pairs, while $f(x)$
stands for the point of the range that is paired with the
point x of the domain. For example, the logarithmic
function consists of the pairs $(x, \log x)$; one of these pairs
is $(e, 1)$, and $\log(e) = 1$. The distinction between a function
and its values is not made consistently in most books. It
looks a little odd, although it is both correct and
unambiguous, to speak of "the function log" (of course,
we must specify the domain of the function; here it would
presumably be $(0, \infty)$). This would usually be called "the
function $\log x$." We often want to talk about more
complicated functions, such as the function whose value at
x is $\log(\sin x)$. Here we are forced into a certain amount of
awkwardness if we are to avoid the possibly misleading

"the function $\log(\sin x)$." The same problem arises when we want to talk about functions that are so simple that they have no generally accepted names. The identity function, whose ordered pairs are (x, x) (for x in a specified domain) should not be called "the function x," but may correctly be described as the function I such that $I(x) = x$. The gain in clarity is well worth the loss of brevity.

If the domain of a function f is in R_1, its values are usually written $f(x)$. If the domain is in R_2, the values are usually written $f(x, y)$, although $f((x, y))$ would be more in accordance with our general principles. The elements of a sequence are conventionally written as s_n rather than as $s(n)$, and we ought to refer to the sequence as s. However, as we have already remarked, it is usually more convenient to call the sequence $\{s_n\}$, that is, to specify what its elements are (since its domain is understood to be the set of positive integers). Thus the sequence whose ordered pairs are $(1, 1), (2, 4), (3, 9), \ldots$ is usually written $\{n^2\}$, and restrictions of it are clearly indicated by $\{n^2\}_{n=3}^{\infty}$, $\{n^2\}_{n \text{ odd}}$, $\{n^2\}_{n=3}^{8}$, and the like. Similarly we could describe the function f such that $f(x) = \sin 2x$ (with domain R_1) as $\{\sin 2x\}$, reserving the notation $\sin 2x$ for the particular point of R_1 (as range) that is paired with the point x of the domain. The restriction of this function to $(0, 2\pi)$ would be $\{\sin 2x\}_{0 < x < 2\pi}$. Various other notations are in use;[12] it will not be worth while to adopt any of them systematically for the purposes of this book.

Originally a function was thought of as being defined by a formula, but for many years there has been little worry about this aspect of functions. Many functions that are defined in an apparently arbitrary way can be represented by formulas, although only by formulas of a rather elaborate kind. (We must, however, recognize that the simple notation $f(x) = \sin x$ conceals a nontrivial limiting process, which we forget because the function is so familiar.) For example, let a function f be defined by putting $f(x) = 1$ for rational x in R_1, $f(x) = 0$ for irrational x. Then we have

$$f(x) = \lim_{m \to \infty} \lim_{n \to \infty} (\cos m! \pi x)^n.$$

Exercise 12.2. Verify this.

A more complicated example of the same kind is[13]

$$f(x) = \lim_{r \to \infty} \lim_{m \to \infty} \lim_{n \to \infty} \sum_{\nu=0}^{m} \left[1 - \left(\cos\{(\nu!)^r \pi/x\} \right)^{2n} \right],$$

which for positive integers x yields the largest prime factor of x.

The functions that can be represented by only one limiting process starting from continuous functions are rather special (see §18). Indeed, not all functions with domain and range in R_1 can be represented by formulas, at least by formulas starting with continuous functions and involving only a countable number of limiting processes.[14] We shall not attempt to construct an example of this phenomenon here.

Although there are nontrivial, and even interesting, properties of completely general functions with a given domain and range in a given space (cf. p. 147), the most interesting and generally useful properties are possessed only by functions that belong to more or less special classes. In other words, the interesting thing is to impose some special property and see what other properties follow as consequences. From the point of view of the general theory it is only a fortunate accident that the functions that arise naturally in the applications of mathematics are frequently continuous, or differentiable, and so on. However, since such special functions are, in fact, frequently encountered, it is both desirable and interesting to know some of their important properties.

13. Continuous functions. We are going to define what we shall mean by continuous functions and then discuss some of their properties. This is not how the concept of continuous function originally got into mathematics. The term *continuous* came first and then people sought a definition that would fit their intuitive feelings about it. For a real-valued function whose domain is an interval in R_1 (the most familiar case), it was, for example, felt at one time that a continuous function should be defined as one that takes on every value between any two values that it assumes; in other words, as a function such that the image of every interval in its domain is an interval or a point. We call this the *intermediate value property*. It does not, unfortunately, force a function to have all the properties that a continuous function might reasonably be expected to have. For example, the function defined by $f(x) = \sin(1/x)$ for all real x except 0, and $f(0) = 0$, has the intermediate value property but does not impress most people as being continuous at 0. By a rather more elaborate construction we can exhibit a function that has the intermediate value property in every interval, however small, but does not look continuous since it manages to have the intermediate value property simply by taking on every value between 0 and 1 in every interval.

We construct this function as follows.[15] Let x, between 0 and 1, be expanded as an ordinary decimal $x = 0.a_1a_2\ldots$, and consider the number $z = 0.a_1a_3a_5\ldots$. If z is not a periodic decimal, put $f(x) = 0$. If, however, z is periodic, with its first period beginning with a_{2n-1}, put

$$f(x) = 0.a_{2n}a_{2n+2}a_{2n+4} \cdots .$$

This defines the required function f. For, if an interval I is given, we can find n so large that I contains a terminating decimal $0.a_1a_2 \ldots a_{2n-1}$ and all the numbers

$$0.a_1a_2a_3 \ldots a_{2n-1}a_{2n} \cdots$$

starting with the same first $2n - 1$ digits. Now let $y = 0.b_1b_2 \ldots$ be any number in $(0, 1)$. We can arrange that $0.a_1a_3 \ldots a_{2n-1}a_{2n+1}a_{2n+3} \ldots$ is periodic with its first period starting at a_{2n-1}, and then according to our construction

$$x = 0.a_1a_2 \ldots a_{2n-1}b_1a_{2n+1}b_2a_{2n+3} \cdots$$

has $f(x) = y$.

It is interesting in this connection that every function from an interval in R_1 to R_1 can be written as the sum of two functions, each of which takes on every real value in every subinterval.[15a]

For functions from an interval in R_1 into R_1, the definition of continuity that has come to be accepted is probably familiar to the reader. We say that f is continuous at x_0 if, given any positive ϵ, we can find a positive number δ such that, if $|x - x_0| < \delta$, then

$$|f(x) - f(x_0)| < \epsilon;$$

and we say that f is continuous in an interval if it is continuous at every point of the interval. The intuitive idea behind this definition is that a small change in the position of the point that we are looking at in the domain should produce a small change in the position of the image point in the range. It must be conceded that this definition still does not correspond as closely as we might wish to the intuitive idea of a continuous function. For example, we cannot necessarily draw a satisfactory graph of a given continuous function on paper with a pencil: think, for example, of the everywhere oscillating functions of §10. Indeed, the intuitive notion of a continuous function is more nearly that of a continuous function whose graph is made up of a finite number of increasing or decreasing pieces.

The definition of continuity can be carried over to the case where domain and range are in any two metric spaces. In the simplest case the domain of f contains a neighborhood of the point x_0; then we say that f is continuous at x_0 if, given any positive number ϵ, we can find a positive number δ such that if $d(x, x_0) < \delta$ then $d(f(x), f(x_0)) < \epsilon$. (Of course, the two d's refer, in general, to different metrics.)

If we want to consider more general situations we must face up to the fact that a function may be continuous or not according to the space in which its domain is supposed to lie. For a simple example, consider a constant function with domain R_1. This function is certainly continuous at each point of its domain. However, if R_1 is regarded as a subset of R_2, we cannot say that the function is continuous at any point, since it is not even defined in any neighborhood in R_2. Indeed, this function is the restriction to R_1 of many functions with domains in R_2; some of these functions are continuous and some are not.

We can always take the domain of a given function as a metric space in itself, and ask whether the function is then continuous; we may add, "with respect to its domain," for emphasis. A new idea enters if we want to consider, not a given function, but its restriction to a subset of its domain.

It may happen that the restriction is continuous (with respect to its own domain) whereas the original function is not continuous. For example, the function f (p. 78) that has the value 1 for rational points of R_1 and the value 0 for irrational points is clearly discontinuous at each point of R_1. The restriction of the same function to the set P of rational points of R_1 is a function g that is continuous (on P as space). Some authors say that the original function is discontinuous at every point of P, but continuous on P with respect to P. The confusion created by statements of this kind can best be avoided by recognizing that to define a function we must say what its domain is as well as how its values are to be calculated. Care in specifying the domain is especially important when the function is defined by a formula.

To say that a function f is continuous at a point x_0 of a set E, with respect to E, then, is to say that the restriction of f to E is continuous at x_0 with E as space. An equivalent definition is obtained by using the (ϵ, δ) definition (p. 80) with the additional requirement that $x \in E$. In particular, let f be a function whose domain is an interval in R_1 containing x_0 in its interior, and let g be the restriction of f to an interval $[x_0, b)$ to the right of x_0. If g is continuous at x_0, it is often said that f is continuous on the right at x_0. This is the same as saying that f satisfies the definition of continuity at x_0 except that only right-hand neighborhoods of x_0 are considered. To have an illustration, let us define functions f_1, f_2, f_3, respectively, all to have the value -1 for $x < 0$ and the value $+1$ for $x > 0$, whereas $f_1(0) = 1$, $f_2(0) = -1$, and $f_3(0) = 0$. Then f_1 is continuous on the right at 0, f_2 is continuous on the left, f_3 is continuous on neither side, and all three functions are discontinuous at 0.

In this connection it is interesting that for any real-valued function whatsoever (with domain an interval) there is a dense (but countable) set E such that f, restricted to E, is continuous on E.[15b] On the other hand, there are functions whose restrictions to all sets of the cardinal number of R_1 are discontinuous.[15c]

Exercise 13.1. Show that if y is a point of a metric space, the function f defined by $f(x) = d(x, y)$ is continuous on the space.

Exercise 13.2. Let E be a closed set in a metric space; let D be the function such that $D(x)$ is, for each point x in the space, the distance (cf. Exercise 8.9) from x to E. Show that D is continuous.

If we want to consider continuous functions on spaces that are not metric, a definition of continuity in terms of distance naturally will not do. Although we shall consider only metric spaces in this book, we shall rephrase the definition of continuity in a form that can be extended to more general spaces, if only because it is often a convenient form to use even in metric spaces. This more sophisticated definition reads: *f is continuous on its domain if and only if the inverse image of each open set in the range space is an open set in the domain.* (Here the domain of f is to be regarded as the space with respect to which open sets are defined.) The inverse image of a set E means, of course, the set of points of the domain whose image points are in E. For example, if $f(x) = \sin x$ with domain R_1, the inverse image of the interval $(0, 1)$ consists of the union of the intervals $(0, \pi), (2\pi, 3\pi), (-2\pi, -\pi), \ldots$, which is an open set. If, however, $f(x) = 1$ for $x > 0$, $f(0) = 0$, and $f(x) = -1$ for $x < 0$, the inverse image of the open interval $(-\frac{1}{2}, \frac{1}{2})$ contains the single point 0, and so is not open.

Exercise 13.3. Give an example to show that the image of an open set under a continuous function is not necessarily open.

To verify the equivalence of the two definitions of continuity in a metric space, suppose first that f is continuous under the original definition. Let E be an open set in the range space and let x_0 be a point in the inverse image of E. Then $f(x_0) \in E$, and if ϵ is small enough, every y such that $d(f(x_0), y) < \epsilon$ belongs to E (since E is open). Since f is continuous, there is a positive δ such that $d(x, x_0) < \delta$ implies $d(f(x), f(x_0)) < \epsilon$. Hence all points x sufficiently near x_0 have their images in E; that is to say, the inverse image of E contains a neighborhood of each of its points, so it is open. Conversely, let the inverse image of every open set be open. In particular, the inverse image of any neighborhood in the range space, defined by the inequality $d(f(x), f(x_0)) < \epsilon$, is open and so contains a neighborhood $d(x, x_0) < \delta$; this is an appropriate neighborhood to associate with ϵ in the original definition.

We have proved a little more than we set out to do, namely that *f is continuous at x_0 if and only if the inverse image of every open set that contains $f(x_0)$ contains a neighborhood of x_0.*

Exercise 13.4. Show that if the range of f is in R_1 and f is continuous at x_0 and $f(x_0) \neq 0$ then there is a neighborhood of x_0 in which $|f(x)|$ has a positive lower bound, that is, $|f(x)| \geqslant m > 0$.

Exercise 13.5. Show that if f has an interval in R_1 as its domain, has its range in R_1, and is continuous at x_0, then f is bounded in some neighborhood of x_0.

Exercise 13.6. If f, from R_1 into R_1, is discontinuous at x_0, there are a sequence $\{x_n\}$ with limit x_0 and a positive ϵ such that $|f(x_n) - f(x_0)| > \epsilon$.

14. Properties of continuous functions. It is a commonplace that sums, products, and quotients of continuous functions are continuous (provided that the divisor in the quotient is not zero). More precisely, we should suppose that the two functions f and g have the same domain, and that their values are in R_1. Then we can define $f + g$, fg, and f/g in the usual (and natural) way, provided in the last case that g is not zero anywhere on the common domain of the functions. Then if f and g are both continuous at x_0, so are $f + g$, fg, and f/g if it is defined. However, it is usually not necessary to be so meticulous. When f and g have different domains, we are likely to write $f + g$ for the sum of their restrictions to the intersection of the domains of f and g, and similarly f/g for the quotient of the restrictions of f and g to the part of the intersection of their domains where g does not take the value 0. Note in this connection that the function f_1 defined by $f_1(x) = x/x$ and the function f_2 defined by $f_2(x) = 1$ are different functions since their domains are different; we cannot say that f_1 is continuous at 0. In discussing the continuity of a product it is convenient to use Exercise 13.5.

Exercise 14.1. If f is continuous at x_0 and g is not, show that $f + g$ is not. Can $f + g$ be continuous at x_0 if neither f nor g is continuous at x_0?

Exercise 14.2. Carry out the detailed proofs for the continuity of $f + g$, fg, and f/g when f and g are continuous.

A function f is said to be *univalent*, or *one-to-one*, if in the set of ordered pairs that constitute the function, not only does no x in the domain occur twice but also no y in the range occurs twice. In this case the ordered pairs (y, x) with y in the range and x in the domain also constitute a function, the *inverse* of f, often denoted by f^{-1}; its domain is the range of f and its range is the domain of f. It is often useful to know that *under certain circumstances the inverse of a continuous univalent function is continuous.* This statement is true *if the domain of the function is a compact set in some* R_n, or, more generally, whenever the domain has the property that every infinite subset of it has a limit point (the conclusion of the Bolzano-Weierstrass theorem).

To verify the preceding assertion, we have to show that the images of open sets are open, since these will be the inverse images of open sets under f^{-1}. It is an equivalent statement that the images of closed sets are closed, a statement which is somewhat more convenient for us to use.

Exercise 14.3. Establish the equivalence asserted in the preceding sentence.

Let, then, E be a closed set in the domain of f, let F be the image of E, and let y_0 be a limit point of F; we have to show that $y_0 \in F$. Let $\{y_n\}$ be a sequence of distinct points of F with $y_n \to y_0$, and let $y_n = f(x_n)$. There is, by univalence, just one x_n for each y_n, and the x_n are all different since the y_n are all different. Then the *set* whose points are the x_n has a limit point x_0 and a subsequence $\{x_{n_k}\}$ has x_0 as its limit. (See p. 57.) We have $x_0 \in E$ since E is closed. Since f is continuous, $y_{n_k} = f(x_{n_k}) \to f(x_0)$; but $y_{n_k} \to y_0$, so $y_0 = f(x_0) \in F$.

Although the intermediate value property discussed in §13 does not characterize continuous functions, it is a property that (under some conditions) continuous functions have. It is possessed by a continuous function with values in R_1 if its domain is a connected set in a metric space S. To see this, let $f(a) = A$ and $f(b) = B$, and let $A < C < B$. Consider the sets E_1 and E_2 in S, consisting respectively of the points x of S for which $f(x) < C$ and of the points x of S for which $f(x) > C$. These sets are disjoint, since $f(x)$ cannot (for the same x) be simultaneously less than C and greater than C. They are not empty, since $a \in E_1$ and $b \in E_2$. They are open, since they are inverse images of open sets. Since the domain of f was assumed connected it cannot be the union of two disjoint nonempty open sets. Hence the domain of f must contain at least one point c that is in neither E_1 nor E_2. The only possible value for $f(c)$ is C.

We showed in §7 that a continuous function whose domain is compact and whose range is in R_1 has a largest and a smallest value.

Exercise 14.4.   A more elegant proof of the preceding statement can be given by supposing that the least upper bound M of the values of f is not a value of f, and considering $1/[M - f(x)]$.

Exercise 14.5.   Deduce that if the domain is both compact and connected, the range of a continuous function with values in R_1 is a closed bounded interval or a point.

Compactness is of course not necessary for the existence of a maximum.

Exercise 14.5a.   Let f be a nonnegative continuous function with domain $[a, \infty)$ and range in R_1, and let $f(x) \to 0$ as $x \to \infty$. Then f has a maximum on $[a, \infty)$.

Exercise 14.5b.   If f is as in Exercise 14.5a, but strictly positive, there is a sequence $\{x_n\}, x_n \to \infty$, such that $f(x) < f(x_n)$ for all $x > x_n$. That is, as $x \to \infty$ the function f never again takes as large a value as it had at x_n.

We now give some applications of the intermediate value property.

Consider a function f, from an interval in R_1 into R_1, that has the intermediate value property on every interval in its domain and has a discontinuity at the point c. Then

(Exercise 13.6) there is a sequence $\{x_n\}$ with limit c such that either $f(x_n) > f(c) + \epsilon$ with some positive ϵ, or $f(x_n) < f(c) - \epsilon$; say the former. Since f has the intermediate value property on every interval, it takes every value between $f(c)$ and $f(c) + \frac{1}{2}\epsilon$. Furthermore it does this infinitely often, since we can always consider a smaller neighborhood of c. Hence *if a function from an interval to R_1 has the intermediate value property on every subinterval and is discontinuous, it must take some values infinitely often.* (Thus the discontinuous function with the intermediate value property, constructed on pp. 79–80, illustrates the typical situation better than one might have expected.) As a corollary, we have that *if a function has the intermediate value property on every interval and takes no value more than once, it is continuous.* In particular, a function is continuous if it takes each value between $f(a)$ and $f(b)$ exactly once, in every interval $[a, b]$ in its domain.[15d]

A strictly monotonic function (p. 148) is an example of a function that takes no value more than once, but not every function with this property is strictly monotonic. (Consider $f(x) = x + 1$ for $-1 < x \leqslant 0$, $f(x) = x - 1$ for $0 < x < 1$.) However, a continuous function that takes no value more than once is indeed strictly monotonic. For, if it were not strictly monotonic there would have to be points $x_1 < x_2 < x_3$ with $f(x_1) \leqslant f(x_2), f(x_3) \leqslant f(x_2)$ (or the corresponding inequalities reversed), and so f would have a maximum at some point between x_1 and x_3 (because f is continuous), and the maximum would be proper (since f takes no value more than once). Then $f(x) < f(c)$ for values of x on both sides of c, and so by the intermediate value property f would take some value near $f(c)$ twice, contradicting the hypothesis. (The same reasoning shows that a continuous function is monotonic between its consecutive maxima and minima: that is, if f

has a maximum at x_1 and a minimum at $x_2 > x_1$, and neither a maximum nor a minimum in (x_1, x_2), then f is monotonic in (x_1, x_2).)

We have just seen that the continuous functions (on an interval in R_1) that take each value exactly once are strictly monotonic. What sort of continuous functions take each value exactly twice? The answer is that there are no such functions.[15e]

In fact, it is just as easy to show that no continuous function on a closed bounded interval can take each of its values exactly n times, $n > 1$. Suppose that there is such a function f. Then f has an absolute maximum and an absolute minimum, and each must be attained at interior points except in the case $n = 2$, when the minimum (say) might be attained at both endpoints. Hence we may suppose that the maximum is attained at an interior point c_1. There are points $c_2, c_3, \ldots, c_n$ where $f(c_k) = f(c_1)$, $k = 2, \ldots, n$, and $f(x) < f(c_1)$ in a deleted neighborhood of each c_k (one-sided if c_k is an endpoint). For a sufficiently small positive ϵ, the line $y = f(c_1) - \epsilon$ intersects the graph of f twice near c_1 and once near each other c_k, by the intermediate value theorem, so there is a value that f takes $n + 1$ times, a contradiction.

Since f cannot be continuous and exactly two-to-one, let us suppose now that f takes each value *at most* twice. Then we may conclude that the graph of f falls into at most three monotonic pieces[15f] (like the graph on p. 93, with a small piece removed at one end). It adds interest to this theorem to observe that nothing similar holds for continuous functions that take each value at most three times: the graph of such a function does not have to be composed of a finite number of monotonic pieces,[15g] as is indicated by the sketch on p. 90, where the nth square from the right has side a_n and $\sum a_n < \infty$.

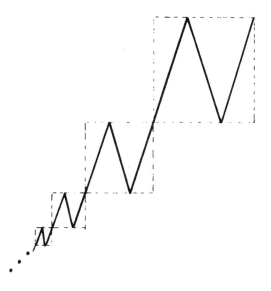

Let us suppose then that f is continuous on $[a, b]$ and at most two-to-one. By the principle mentioned above, f is monotonic between any two consecutive maxima and minima. Thus if maxima and minima of f are attained only at the endpoints, f is itself monotonic. If it is not monotonic it has at least one maximum or minimum inside the interval, say a maximum at c. If f has no other interior maxima or minima, it must be monotonic on (a, c) and on (c, b). The next case is the one where f has precisely one interior maximum at c_1 and one interior minimum at c_2, say with $a < c_1 < c_2 < b$. Then f is monotonic on each of the three intervals $(a, c_1), (c_1, c_2)$, (c_2, b). Finally suppose that f has more than a total of two interior maxima and minima. Let a maximum, a minimum, and a maximum (not necessarily consecutive) occur at c_1, c_2, c_3, where $c_1 < c_2 < c_3$ and $f(c_1) > f(c_2), f(c_3)$ $> f(c_2)$; suppose for definiteness $f(c_1) \geqslant f(c_3)$. Then f takes some value slightly less than $f(c_3)$ at least twice; taking this value to be greater than $f(c_2)$, it will be less than $f(c_1)$, and by the intermediate value property it will be taken again between c_1 and c_2, for a total of three times, contradicting the hypothesis.

We turn now to a different application of the intermediate value property. Let f be a continuous function from an interval in R_1 into R_1. We say that f has a *horizontal chord* of length a if there is a point x such that x and $x + a$ are both in the domain of f and $f(x) = f(x + a)$. This means that there is a horizontal line segment of length a having both ends on the graph of the function; we do not care whether or not the segment has other points in common with the graph. For example, if $f(x) = 1$ for every x, f has horizontal chords of all lengths; the segment from $(-1, 1)$ to $(1, 1)$ is a horizontal chord of length 2 of the function f defined by $f(x) = x^3 - x + 1$; the function f defined by $f(x) = x^3$ has no horizontal chords.

A function f whose domain is all of R_1 is said to be *periodic* with period p if $f(x + p) = f(x)$ for all x.

Exercise 14.6. Show that a continuous periodic function is bounded.

Exercise 14.7. Show that a continuous periodic function has a maximum.

Exercise 14.8. Show that if f is continuous and has period p then

$$\int_x^{x+p} f(t)\, dt$$

has a value independent of x.

We first notice that *a continuous periodic function f has horizontal chords of all lengths*. That is, if f has period p, and a is any real number, there is an x such that $f(x + a) - f(x) = 0$.

To see this, consider

$$\int_0^p \left[f(x + a) - f(x) \right] dx,$$

which is zero (by Exercise 14.8). Hence the integrand must change sign (unless it vanishes identically; then $f(x + a) \equiv f(x)$ and there is no more to be said). But the integrand has period p, and if it changes sign once in a period it must change sign at least twice in order to return at p to the value it had at 0, unless it was 0 at 0 to start with. Hence there are at least two points x in $[0, p)$ for which $f(x + a) = f(x)$, and we see that *a continuous function of period p has two horizontal chords of any given length, with their left-hand endpoints at different points of $[0, p)$.*[15h]

Exercise 14.9. Show that a periodic continuous function always has a chord (not necessarily horizontal), of prescribed span, with its midpoint on the graph: that is, for each a there is an x such that $f(x + a) - f(x) = f(x) - f(x - a)$.

For functions that are not periodic the situation is quite different. A given continuous function f, say with domain $[0, 1]$, may have no horizontal chords at all. However, let us suppose to begin with that it does have one horizontal chord. More specifically, *suppose that $f(0) = f(1)$, so that the segment $[0, 1]$ is a horizontal chord.* The universal chord theorem[16] states that *there are then horizontal chords of lengths $\frac{1}{2}, \frac{1}{3}, \frac{1}{4}, \ldots$, but not necessarily a horizontal chord of any given length that is not the reciprocal of an integer.*

To prove the positive half of this theorem, let k be a positive integer and consider the continuous function g defined by $g(x) = f(x + 1/k) - f(x)$, whose domain is $[0, 1 - k^{-1}]$. Then we assert that 0 is in the range of g. If this is not the case, g would be either positive for all x in its domain or else negative for all x in its domain (by the intermediate value property), and so $g(0) + g(1/k) + g(2/k) + \cdots + g(1 - 1/k)$ would be either positive or negative; on the other hand, this sum "telescopes" and is equal to $f(1) - f(0) = 0$.

Alternatively, if $g(x) > 0$ for all x, that is, if $f(x) < f(x + 1/k)$ for all x, then $f(0) < f(1/k) < f(2/k) < \cdots < f((k-1)/k) < f(k/k) = f(1) = f(0)$, a contradiction.

The graph below indicates a function f that has a horizontal chord of length 1 but no horizontal chords of length a for $\frac{1}{2} < a < 1$. Naturally a function of this kind has horizontal chords of *some* lengths that are not reciprocals of integers; the negative part of the universal chord theorem asserts that, for each number b that is not the reciprocal of an integer, there is a continuous function that has a horizontal chord of length 1 but does not have a horizontal chord of the particular length b.

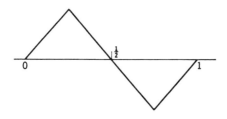

Exercise 14.10. Show how to construct an example for the negative part of the universal chord theorem for every $b \neq 1/k$. (The symmetry of the example given for $\frac{1}{2} < b < 1$ is misleading.)

An interesting complement to the universal chord theorem is that[17] *a continuous function f that has a horizontal chord of length* 1 *always has either a horizontal* *chord of length a or two different ones of length* $1 - a$ (if $0 < a < 1$). To see this, suppose that $f(0) = f(1)$ and form a new function g by repeating f with period 1. As a periodic continuous function, g necessarily has two horizontal chords of length a with left-hand endpoints in $[0, 1)$ (see p. 92). A horizontal chord of length a for g, starting at x, is a horizontal chord for f unless $x + a > 1$. If $x + a > 1$, $0 < x + a - 1 < 1$ and so a horizontal chord of length $1 - a$ for g (also for f) starts at $x + a - 1$ and ends at x.

Exercise 14.11. Prove the universal chord theorem by induction, starting from the fact that if $f(0) = f(1)$, then f has either a horizontal chord of length a or one of length $1 - a$.

In the same way we can show that if f has a derivative f' in $[0, 1]$ and $f'(0) = f'(1)$ then for any integer n there are points x and $x + n^{-1}$ such that f has the same slope at both points. This depends on the fact (p. 136) that a derivative always has the intermediate value property; this property (for $g(x)$, not merely for $f(x)$) was all that was used in the proof of the universal chord theorem.

Another application of the same idea yields a simple fixed-point theorem. This says that *a continuous mapping of an interval into (part or all of) itself has at least one fixed point*; that is, there is at least one point that coincides with its image. Another way of saying this is that if f is a continuous function on $[0, 1]$, such that $0 \leqslant f(x) \leqslant 1$, then the graph of the function must cross the line $y = x$.

Exercise 14.12. Prove this; that is, prove that if f is continuous in $[a, b]$ and all the values of f are in $[a, b]$, then there is an x in $[a, b]$ for which $f(x) = x$.

Exercise 14.12a. Suppose that a clock runs irregularly but that at the end of 24 hours it has neither gained nor lost overall. Is there some hour for which this clock shows an elapsed time of exactly one hour? Is there some continuous 576 minutes during which it shows an elapsed time of 576 minutes? (Assume that the indicated time is a continuous increasing function.)[17a]

Another closely related theorem states that if a continuous periodic function, of period $2p$, has the property that $f(x) = -f(x + p)$ for every x (like the graph of $y = \sin x$, with $p = \pi$), then $f(x) = 0$ for at least one x. This is obvious from the intermediate value theorem. If formulated in a different way, it becomes a simple case of Borsuk's antipodal point theorem, which is much harder to prove in higher-dimensional situations.[18] In this formulation we consider a continuous function f whose domain is the circumference of a circle in R_2 and whose range is in R_1. Suppose that the image of every pair of antipodal points (points at opposite ends of a diameter) is a pair of points that are symmetric with respect to the origin. Then some point of the circumference maps into the origin.

Still another application of the intermediate value theorem shows that a (two-dimensional) pancake of arbitrary shape can be bisected by a knife cut in any specified direction. At least if the boundary of the pancake is sufficiently simple, it is easy to see that the part of the pancake lying on one side of a line in the given direction has an area that varies continuously as the line is moved parallel to itself. Since this area can be either 0 or the total area of the pancake it must at some time be exactly half the total area.

We now indicate how two pancakes in a plane can be bisected simultaneously by some line. Let the two pancakes be A and B. As we have just seen, there is a line in any given direction bisecting B. For the direction determined by any point P on a given circle in R_2, with center O, find a line bisecting B and let $f(P)$ be the signed difference between the area of the part of A on the left of

this line and the area of the part of *A* on the right. We
have a function whose domain is a circumference; whose
range is in R_1; and which maps antipodal points into
points symmetric about *O*, since replacing *P* by its
antipode interchanges left and right. If we can show that *f*
is continuous, the special case of Borsuk's theorem, noted
above, shows that $f(P) = 0$ for some *P*, which is to say
that a line in the direction *OP* bisects both *A* and *B*. The
continuity of *f* is plausible, but not quite obvious. It is
enough to show that a small change in the position of *P*
produces not only a small change in the direction of the
line bisecting *B*, but also a small change in its position, for
example, a small change in its intercept on one of the
coordinate axes. For, if this statement is true, $f(P)$ will
change by only a small amount. Now the preceding
statement about lines bisecting *B* is true without reservation only if we impose some restriction on the admissible
shape of a pancake; for example, a "pancake" shaped like
○—○ can be bisected by many vertical lines. If we
restrict ourselves to convex pancakes there is no difficulty,
as is clear from a figure.[18a]

Exercise 14.13. Given a convex closed curve in the plane, there
is a line that simultaneously bisects the curve and the area that it
surrounds.[19]

There are similar theorems in more dimensions but they
are harder to prove. The three-dimensional fixed-point
theorem can be stated in picturesque, if misleading,
language: if a cup of coffee is stirred in any continuous
fashion, there is at least one molecule that ends up in its
original position. (This is correct only if the coffee
occupies every point inside the cup and "molecule" is
interpreted as "point.") The three-dimensional antipodal

point theorem states that a continuous mapping of the surface of a sphere into the space R_2, which carries every pair of antipodal points into a pair of points symmetric with respect to the origin, must carry some point of the surface of the sphere into the origin. This can be used to show that any three volumes in space can be bisected simultaneously by some plane (the "ham sandwich" theorem).[20]

15. Upper and lower limits. We shall need to use a generalization of the notion of limit of a sequence of real numbers. If $\{s_n\}$ is such a sequence, and the numbers s_n form a bounded set, we shall show that there is always a number L with the following property: Given any positive ϵ, we have $s_n \leqslant L + \epsilon$ whenever n is sufficiently large, and in addition an infinite number of s_n satisfy $s_n \geqslant L - \epsilon$. This number L is called the *upper limit* of $\{s_n\}$, and is written $\limsup s_n$ or $\overline{\lim} s_n$. If $\{s_n\}$ is unbounded above we write $\limsup s_n = +\infty$. If $\{s_n\}$ is unbounded below, L may fail to exist, and in this case we write $\limsup s_n = -\infty$.

Consider some examples. (i) Let $s_n = (-1)^n$, so that our sequence is $1, -1, 1, -1, \ldots$. Then $\limsup s_n = 1$. (ii) Let $s_n = n$; the sequence is $1, 2, 3, \ldots$. Then we have $\limsup s_n = +\infty$. (iii) Let $\{s_n\} = \{-1, -2, \ldots\}$. Then $\limsup s_n = -\infty$. (iv) Let $s_n = 1/n$, $n = 1, 2, 3, \ldots$. Then $\limsup s_n = \lim s_n = 0$. (v) Let

$$\{s_n\} = \left\{\frac{1}{2}, \frac{1}{3}, \frac{2}{3}, \frac{1}{4}, \frac{3}{4}, \frac{1}{5}, \frac{4}{5}, \cdots \right\};$$

then $\limsup s_n = 1$.

Exercise 15.1. By considering the numbers L_n defined by $L_n = \sup s_k$ for $k \geqslant n$, show that $\limsup s_n$ exists if $\{s_n\}$ is bounded.

Exercise 15.2. Define the lower limit, lim inf, or $\underline{\lim}$, similarly, and determine it for the five examples given above for lim sup.

The definition of upper limit resembles that of least upper bound, from which it differs in that finite subsets of the elements s_n are disregarded. For example, the value of $\limsup s_n$ is unchanged if the first thousand s_n are replaced by other numbers. We could also define $\limsup s_n$ as the largest limit obtainable by picking convergent subsequences out of $\{s_n\}$; it is this definition that accounts for the terminology.

We have $\limsup(s_n + t_n) \leqslant \limsup s_n + \limsup t_n$ when both the quantities on the right are finite; but strict inequality may occur here, for example, if $\{s_n\} = \{1, 0, 1, 0, \ldots\}$ and $\{t_n\} = \{0, 1, 0, 1, \ldots\}$. However, we have $\limsup(s_n + t_n) = \limsup s_n + \lim t_n$ if $\lim t_n$ exists.

Exercise 15.3. Prove the inequality and equality stated above.

The inequality extends to finite sums, but not to infinite sums. For example, if we let s_k denote the sequence $\{0, 0, \ldots, 0, 1, 0, \ldots\}$ with elements $s_{k,n}$ equal to 0, except that the kth element $s_{k,k}$ of the kth sequence is 1, we have (as $n \to \infty$) $\limsup(s_{1,n} + s_{2,n} + \cdots) = \limsup 1 = 1$, whereas $\limsup s_{k,n} = 0$ for each k.

Upper and lower limits can also be defined for functions whose ranges are in R_1 but whose domains are more general. If the domain of f is a set S in a metric space and x_0 is a limit point of S, by $\limsup_{x \to x_0} f(x) = L$ we mean that given any positive ϵ we have $f(x) \leqslant L + \epsilon$ whenever x is in a sufficiently small neighborhood of x_0, and in addition there is a sequence of points $\{x_n\}$ with limit x_0 such that $f(x_n) \geqslant L - \epsilon$. Slight modifications have to be made if $L = \pm\infty$. In a similar way we can define

$\limsup_{x \to +\infty} f(x)$ if the domain of f is a subset of R_1 that is unbounded above. Examples: (i) If $f(x) = \sin x$ for $x \in R_1$, we have

$$\limsup_{x \to +\infty} f(x) = +1 \quad \text{and} \quad \liminf_{x \to +\infty} f(x) = -1.$$

(ii) If $f(x) = e^x \sin x$ for $x > 0$, then

$$\limsup_{x \to +\infty} f(x) = +\infty \quad \text{and} \quad \liminf_{x \to +\infty} f(x) = -\infty.$$

(iii) If $f(x) = e^{1/x}$ for $x \neq 0$, we have $\limsup_{x \to 0+} f(x) = +\infty$ and $\liminf_{x \to 0-} f(x) = 0$.

Exercise 15.4. Show that if $\limsup s_n = \liminf s_n = L$ and L is finite, then $\lim s_n = L$ according to the definition given in §8.

Exercise 15.5. If $\limsup s_n \leqslant L$ and $\liminf s_n \geqslant L$, then $\lim s_n$ exists (and equals L).

Up to this point, although we have considered limits of sequences, we have not had to consider limits of more general functions. For functions with values in R_1 it is simplest to define $\lim_{x \to x_0} f(x)$ to be the common value of $\limsup_{x \to x_0} f(x)$ and $\liminf_{x \to x_0} f(x)$. If the domain of f includes a right-hand neighborhood of x_0, we define as $\limsup_{x \to x_0+} f(x)$ the upper limit of the restriction of f to a right-hand neighborhood of x_0: similarly for lim inf. The common value (if any) of these right-hand upper and lower limits is denoted by $\lim_{x \to x_0+} f(x)$, or more compactly by $f(x_0 +)$. There is a similar definition for $f(x_0 -)$.[20a]

16. Sequences of functions. Before we go on to consider the special properties of further classes of functions it will be convenient to discuss several kinds of convergence of sequences of functions.

Suppose that we have a sequence whose elements are functions s_n with a common domain and with values $s_n(x)$ belonging to R_1. We may fix our attention either on the sequence $\{s_n\}$, whose elements are the functions themselves, or on the sequences $\{s_n(x)\}$ whose elements are the values of the functions at the individual points x of the domain. We say that $\{s_n\}$ *converges pointwise* on a set S if the sequences $\{s_n(x)\}$ of real numbers converge for each x in S. For example, let $s_n(x) = x^n$, with $0 \leqslant x \leqslant 1$. For each x in $[0, 1]$, the sequence $\{s_n(x)\}$ converges; the limit defines the discontinuous function L such that $L(x) = 0$ for $0 \leqslant x < 1$, but $L(1) = 1$. On the other hand, we may consider the same functions s_n as points of the space C. Then the sequence $\{s_n\}$ does not converge; indeed, $d(s_n, s_{2n}) = \max_{0 \leqslant x \leqslant 1}(x^n - x^{2n})$, and if we take $x = 2^{-1/n}$ we see that we certainly have $d(s_n, s_{2n}) \geqslant x^n - x^{2n} = \frac{1}{4}$. Hence $\{s_n\}$ cannot be a Cauchy sequence.

Exercise 16.1. Examine $\{s_n\}$ for convergence in C, (a) if $s_n(x) = x^n(1 - x)$, (b) if $s_n(x) = nx^n(1 - x)$; in each case $0 \leqslant x \leqslant 1$.

We see that a sequence of continuous functions can converge pointwise even though the same sequence of functions, considered as elements of the space C, does not converge. It is clear that convergence of a sequence of elements of C does imply the pointwise convergence of the corresponding sequence of functions. There are, however, other metric spaces whose elements are functions for which convergence of a sequence of elements of the space does not guarantee the pointwise convergence of the sequence of functions. An example is the space, mentioned in §4, whose elements are continuous functions on $[0, 1]$, with metric given by

$$d(x, y) = \left\{ \int_0^1 |x(t) - y(t)|^2 dt \right\}^{1/2}$$

Consider a sequence of elements of this space defined as follows: if $2^n \leqslant k < 2^{n+1}$, $x_k(t) = 0$ except in the interval $(2^{-n}k - 1, 2^{-n}(k + 1) - 1)$; in this interval, the graph of x_k is a triangle with base of length 2^{-n} and altitude 1; this triangle moves back and forth as k increases, thus preventing pointwise convergence; but $d(x_k, 0) < 2^{-n/2} \rightarrow 0$.

Whether or not a sequence of functions converges pointwise, if the functions have a common domain, and range in R_1, we can always define pointwise upper and lower limits, $\limsup s_n(x)$ and $\liminf s_n(x)$. If these upper and lower limits are finite, they define two functions, which we can call $\limsup s_n$ and $\liminf s_n$.

It is often convenient to generalize the space C of continuous functions by considering continuous functions whose domains are sets more general than intervals in R_1. In defining C, we used the existence of the maximum of a continuous function whose domain is a compact interval. Since every continuous function whose domain is a compact set has a maximum, we can use just the same definition: let E be a compact set in a metric space; then the space C_E consists of continuous functions f with domain E and range in R_1, with $d(f, g)$ defined by $\max_E |f(x) - g(x)|$. Similarly we can define a space B_E of bounded functions with domain E, by taking $d(f, g)$ defined by $\sup_E |f(x) - g(x)|$; here E does not have to be compact.

If a sequence of continuous functions, regarded as a sequence of elements of C_E, converges, it is customary to say that it *converges uniformly* on E. More generally, suppose that we have a sequence of functions (bounded or not, continuous or not); if the difference of each two of the functions is bounded, we can form the distance between them in the metric of B_E; and if the sequence of functions is a Cauchy sequence in this metric, we say that it converges uniformly on E. For example, if $f(x) = 1/x$ the sequence $\{f, f, f, \ldots\}$ converges uniformly on $0 < x \leqslant 1$. The sequence $\{s_n\}$ with $s_n(x) = x^n$ converges uniformly on any specified interval $[0, a]$ with $0 < a < 1$, but not on $[0, 1]$. Note that it does not converge uniformly on the half-closed interval $[0, 1)$ either. It is important to realize that "uniform convergence on each closed subinterval of an open interval" is not the same as "uniform convergence on the open interval."

Another, more conventional way of saying that the sequence $\{s_n\}$ is uniformly convergent on E is to say that, given a positive ϵ, there is an N such that, for n and m exceeding N, we have, simultaneously for every x in E, $|s_n(x) - s_m(x)| < \epsilon$; we emphasize that N is to be independent of which x in E is being considered. If N is allowed to depend on x, we have the definition of pointwise convergence again.

Suppose that we can find numbers M_n such that $|s_n(x) - s_{n+1}(x)| \leqslant M_n$ for all x in E; that is, we have $\sup_{x \in E} |s_n(x) - s_{n+1}(x)| \leqslant M_n$. Then if ΣM_n converges, the sequence $\{s_n\}$ converges uniformly on E. For, if $m > n$ we have $|s_n(x) - s_m(x)| \leqslant M_n + M_{n+1} + \cdots + M_m$, and the sum on the right is small when m and n are large. This is known as the Weierstrass M-test, and is usually stated in the equivalent form that it assumes when the s_n are regarded as the partial sums of a series. This reads as follows: *If c_n are functions with domain E, and $|c_n(x)| \leqslant M_n$ for all x in E* (where we emphasize that M_n is independent of x), *then* Σc_n *converges uniformly if* ΣM_n *converges*.

It is often useful to consider another kind of convergence: pointwise convergence together with uniform boundedness, that is, boundedness in the metric of B_E. We call this *bounded convergence*. For example, the sequence $\{s_n\}$ with $s_n(x) = x^n$ is boundedly convergent on $[0, 1]$, although it is uniformly convergent only on each $[0, a]$, $0 < a < 1$. If $s_n(x) = nx^n$, we still have $\{s_n\}$ uniformly convergent on each $[0, a]$, $0 < a < 1$, but $\{s_n\}$ is not boundedly convergent on $[0, 1)$.

Exercise 16.2. Show that the limit of a boundedly convergent sequence of functions is a bounded function.

17. Uniform convergence. One of the common uses of uniform convergence arises from the fact that *the limit of a uniformly convergent sequence of continuous functions is continuous*. The same fact can be stated more concisely by saying that *the space C_E is complete*. This is easily proved. In fact, we prove a slightly more general statement that is sometimes useful: *if each s_n is continuous at x_1, and if $\{s_n\}$ converges uniformly in a neighborhood of x_1, then the function L defined by $L(x) = \lim s_n(x)$ is continuous at x_1*. In the first place, uniform convergence implies pointwise convergence, so there is a function L. Let D be the distance function in the space B_E. Then, given a positive ϵ, we have

$$|L(x_1) - L(x_2)| \leqslant |L(x_1) - s_n(x_1)| + |L(x_2) - s_n(x_2)|$$
$$+ |s_n(x_1) - s_n(x_2)|$$
$$\leqslant D(L, s_n) + D(L, s_n)$$
$$+ |s_n(x_1) - s_n(x_2)|.$$

The two D terms on the right can each be made less than $\epsilon/3$ by choosing n large enough, since $\{s_n\}$ converges uniformly. After choosing n large enough, fix n. Then, since each s_n is continuous at x_1, the last term is at most $\epsilon/3$ if $d(x_1, x_2) < \delta$. Hence $d(x_1, x_2) < \delta$ implies that $|L(x_1) - L(x_2)| < \epsilon$ and thus L is continuous at x_1.

Of course, the limit of a nonuniformly convergent sequence of continuous functions is not necessarily discontinuous. For instance, the sequence $\{s_n\}$, where $s_n(x) = nx^n(1 - x)$, is not uniformly convergent on $[0, 1]$ (Exercise 16.1), but its limit is the continuous function 0. However, under additional restrictions it is possible to conclude that if the limit function is continuous the convergence is uniform. For example, *if a sequence of continuous functions converges monotonically to a continuous function on a compact interval in R_1, the convergence is necessarily uniform.*[21] The hypothesis of monotonic convergence means that either $s_n(x) \geqslant s_{n+1}(x)$ for every n and every x in the interval, or $s_n(x) \leqslant s_{n+1}(x)$ for every n and every x in the interval.

Let us take the hypothesis in the form $s_n(x) \geqslant s_{n+1}(x)$. If the limit function is denoted by L, we then have $s_n(x) - L(x) \geqslant 0$. If the convergence is not uniform, $\max_x[s_n(x) - L(x)]$ does not approach zero, so there is a sequence of values of n for which $\max_x[s_n(x) - L(x)] > b > 0$. Since $s_n - L$ is a continuous function it attains its maximum at a point x_n; from the set $\{x_n\}$ we can, by Exercise 8.8, select a sequence $\{y_k\}$ with limit z. Then we have $s_k(y_k) - L(y_k) > b$, and consequently $s_n(y_k) - L(y_k) > b$ for each $n \leqslant k$ (it is only here that we make essential use of the inequality $s_n(x) \geqslant s_{n+1}(x)$). Letting $k \to \infty$, with fixed n, we infer (by the continuity of $s_n - L$) that $s_n(z) - L(z) \geqslant b$ for every n. On the other hand, $s_n(z) - L(z) \to 0$ since we have convergence at the point z. Thus the assumption of nonuniform convergence leads to a contradiction.

Another condition that produces the same result is that the functions s_n are monotonic (even if they are not necessarily continuous). More precisely, *let $s_n \to L$ pointwise on a compact* *interval* $[a, b,]$, *let* L *be continuous, and let all the* s_n *be nondecreasing functions. Then* $s_n \rightarrow L$ *uniformly.*

The proof of this theorem requires some facts from later sections, but since the theorem fits in well here we prove it now. Given a positive number ϵ, choose a finite set of points x_k in $[a, b]$ such that $a = x_1 < x_2 < \cdots < x_m = b$, and $0 \leqslant L(x_k) - L(x_{k-1}) < \epsilon$ for $k = 2, 3, \ldots, m$. Since L is uniformly continuous (see §19) and nondecreasing, these inequalities will certainly hold if the distances between consecutive x_k's are small enough. Moreover, since L is nondecreasing, we have $0 \leqslant L(x_k) - L(x) < \epsilon$ for $x_{k-1} \leqslant x \leqslant x_k$. Now since the sequence of functions $\{s_n\}$ converges pointwise, and there are only a finite number of points x_k, we can choose n so large that $|s_n(x_k) - L(x_k)| < \epsilon$ for all k. Since any x is in some $[x_{k-1}, x_k]$ and L is nondecreasing, we have

$$s_n(x) \leqslant s_n(x_k) \leqslant L(x_k) + \epsilon \leqslant L(x) + 2\epsilon,$$

where we use successively the fact that s_n is nondecreasing, the inequality $s_n(x_k) \leqslant L(x_k) + \epsilon$, and the inequality $L(x_k) \leqslant L(x) + \epsilon$, which follow from what has just been established. Similarly,

$$L(x) - 2\epsilon \leqslant L(x_{k-1}) - \epsilon \leqslant s_n(x_{k-1}) \leqslant s_n(x).$$

These two sets of inequalities together imply that, if n is large enough, $|s_n(x) - L(x)| \leqslant 2\epsilon$ for every x in $[a, b]$; this is the statement of the uniform convergence of $\{s_n\}$.

The limit of a sequence of discontinuous functions may be either continuous or discontinuous, whether or not the convergence is uniform.

One reason for the importance of the idea of uniform convergence is that a good way to construct a function with some special property is often to exhibit it as the uniform limit of functions that do not quite have the property.

As an illustration of this principle (sometimes called the condensation of singularities), we shall exhibit *a continuous curve that passes through every point of a plane area*. (Such area-filling curves are known as Peano curves.) We must of course decide in advance what the phrase "continuous curve" is to mean; the lesson of our construction is that a natural definition of continuous curve may lead to an object that does not fit the intuitive idea of what a continuous curve should look like.[22]

One natural way of defining a continuous curve in R_2 is to say that it is a continuous image of a line segment, that is, the set of values of a continuous function from a convenient closed interval (say $[0, 1]$) to R_2. Of course, different functions may lead to the same image, but this is irrelevant here; we are going to show that there is at least one function for which the image of the interval covers every point of the whole square, indeed, covers some points more than once. We may represent the points p in the image by their coordinates, letting the image point $(x(t), y(t))$ correspond to the point t of the domain. This amounts to saying that we are thinking of a continuous curve as defined by a pair of parametric equations, $x = x(t)$, $y = y(t)$, with continuous functions x and y.

We are now going to construct a continuous curve, in this sense, that passes through every point of the square where $0 \leq x \leq 1$, $0 \leq y \leq 1$. As a matter of fact, the curve that we shall construct passes through some points of the square four times. It is possible to refine the construction so that the curve has nothing worse than triple points, but farther than this we cannot go, as is shown in topology.[23] We shall verify presently that the curve must have double points at least, that is, that there is no continuous one-to-one mapping of a line segment onto a square.

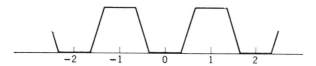

We base our construction[24] on the properties of a continuous function f that is even, is periodic with period 2, has the value zero on $[0, \frac{1}{3}]$, has the value 1 on $[\frac{2}{3}, 1]$, and is linear in $(\frac{1}{3}, \frac{2}{3})$. Define two functions x and y by

$$x(t) = \frac{1}{2} f(t) + \frac{1}{2^2} f(3^2 t) + \frac{1}{2^3} f(3^4 t) + \cdots,$$

$$y(t) = \frac{1}{2} f(3t) + \frac{1}{2^2} f(3^3 t) + \frac{1}{2^3} f(3^5 t) + \cdots.$$

Both series are uniformly convergent (by the M-test), and hence x and y are continuous functions.

Let $0 \leqslant x_0 \leqslant 1$ and $0 \leqslant y_0 \leqslant 1$, and represent x_0 and y_0 by "decimals" in base 2 (cf. p. 39):

$$x_0 = 0.a_0 a_2 a_4 \ldots \text{ (base 2)},$$

$$y_0 = 0.a_1 a_3 a_5 \ldots \text{ (base 2)}.$$

Now define a number t_0 by taking its expansion in base 3 to be $t_0 = 0.(2a_0)(2a_1)(2a_2) \ldots$ (base 3). That is, t_0 is constructed by doubling the binary digits of x_0 and y_0, interlacing them, and interpreting the result in the base 3. We shall now show that $x(t_0) = x_0$ and $y(t_0) = y_0$, so that the curve whose parametric equations are $x = x(t)$, $y = y(t)$ passes through (x_0, y_0).

To do this, we show that $f(3^k t_0) = a_k$ for $k = 0, 1, 2, \ldots$; it will then be obvious from the definition of $x(t_0)$ and $y(t_0)$ that $x(t_0) = x_0$ and $y(t_0) = y_0$. Now a_k is either 0 or 1. If $a_k = 0$, the number represented by $0.(2a_k)(2a_{k+1}) \ldots$

(base 3) is between 0 and $\frac{1}{3}$, and so

$$f(3^k t_0) = f(0.(2a_k)(2a_{k+1}) \dots) = 0;$$

if $a_k = 1$, the number represented by $0.(2a_k)(2a_{k+1}) \dots$ (base 3) is between $\frac{2}{3}$ and 1 and so $f(3^k t_0) = 1$.

We now show that *there cannot be a continuous curve which passes through every point of a square precisely once.* If there were such a curve, it would be the graph of a univalent function with domain an interval in R_1, say $[0, 1]$, and range a square in R_2. By the theorem on pp. 85–86 this function has a continuous inverse. Since both the function and its inverse have the property that the inverse images of open sets are open, it is equally true that the images of open sets are open. Consider the open sets $(0, \frac{1}{2})$ and $(\frac{1}{2}, 1)$ in R_1; their closures have just one point in common. Their images in R_2 are two sets E_1 and E_2 which are also open, and which fill a square except for one point P. If we take a neighborhood in E_1 and a neighborhood in E_2 we can obviously draw a line segment connecting some point of one neighborhood with some point of the other neighborhood, remaining in the square and not going through P. Thus we obtain two disjoint sets, both open, and neither empty, covering a line segment; this contradicts the connectedness of R_1. Hence our hypothetical continuous curve cannot exist.

On the other hand, we showed in §3 that there is a one-to-one correspondence between an interval and a square; in the light of what we have just proved, it cannot be continuous.

A useful theorem that uses the idea of uniform convergence can be stated concisely in the following form: *A uniformly convergent sequence can be integrated term by term.* More precisely, if f_n are functions from a finite real interval I to R_1, if $f_n \to f$ uniformly on I, and if each f_n is integrable in the ordinary (Riemann) sense over I, then

$$\int_I f_n(x)\,dx \to \int_I f(x)\,dx.$$

The proof is immediate if we know that f is integrable: if $I = [a, b]$, we have

$$\left| \int_I f_n(x)\,dx - \int_I f(x)\,dx \right| = \left| \int_I \left[f_n(x) - f(x) \right] dx \right|$$

$$\leqslant \int_I |f_n(x) - f(x)|\,dx$$

$$\leqslant (b - a)\sup_x |f_n(x) - f(x)|$$

$$\to 0.$$

The same proof shows that the sequence of indefinite integrals

$$\int_a^y f_n(x)\,dx$$

converges uniformly itself. If each f_n is, for example, continuous, then f is continuous (p. 103) and so integrable.

In general, to show that the uniform limit of a sequence of integrable functions is integrable demands more of the theory of integration than we shall take up in this book. It is also rather inefficient to prove it in the setting of Riemann integration, since, if we use Lebesgue integrals instead, we have the much more powerful theorem: if each f_n is integrable (in the Lebesgue sense) and $f_n \to f$ boundedly on a finite interval I, then f is integrable and

$$\int_I f_n(x)\,dx \to \int_I f(x)\,dx.$$

Even more than this is true: it is enough to have $|f_n(x)| \leqslant g(x)$ with g integrable on I. (In this case $\{f_n\}$ is said to converge dominatedly.) This is one of the reasons for preferring Lebesgue integration to Riemann integration.

However, even the simplest case, that in which we have a uniformly convergent sequence of continuous functions, has many interesting applications. Here is one of them.

Let f be a function with derivatives of all orders, which then are necessarily all continuous functions (p. 131). Suppose that $\lim_{n\to\infty} f^{(n)} = L$, say, exists uniformly; we might reasonably call L a derivative of f of infinite order. Since on the one hand

$$\int_a^x f^{(n)}(t)\, dt = f^{(n-1)}(x) - f^{(n-1)}(a) \to L(x) - L(a),$$

and on the other,

$$\int_a^x f^{(n)}(t)\, dt \to \int_a^x L(t)\, dt$$

by our theorem on integrating uniformly convergent sequences, we have

$$\int_a^x L(t)\, dt = L(x) - L(a),$$

whence

$$L(x) = L'(x),$$

and so

$$L(x) = ce^x.$$

Thus *any derivative of infinite order is necessarily a simple exponential* [25] (including the case where it is identically zero), no matter what function gave rise to it.

The same theorem on uniform convergence is often used to establish a theorem on termwise differentiation of a sequence of functions. *If the functions f_n have continuous derivatives on an interval I, if $\{f_n(a)\}$ converges for some a in I, and if $\{f_n'\}$ converges uniformly, then f_n converges uniformly to a limit f which is differentiable, and $\lim f_n' = f'$.*

Under the hypotheses we have stated, let $f_n' \to g$. Then

$$f_n(x) - f_n(a) = \int_a^x f_n'(t)\,dt \to \int_a^x g(t)\,dt.$$

Since $f_n(a) \to f(a)$, it follows that $f_n(x)$ converges for each x; indeed,

$$
\begin{aligned}
|f_n(x) - f_m(x)| &= |f_n(x) - f_n(a) - f_m(x) + f_m(a) \\
&\quad + f_n(a) - f_m(a)| \\
&\leqslant |f_n(x) - f_n(a) - f_m(x) + f_m(a)| \\
&\quad + |f_n(a) - f_m(a)| \\
&\to \left| \int_a^x g(t)\,dt - \int_a^x g(t)\,dt \right| = 0,
\end{aligned}
$$

so that $\{f_n\}$ converges uniformly. Now

$$f_n(x) - f_n(a) \to f(x) - f(a),$$

whence

$$f(x) - f(a) = \int_a^x g(t)\,dt.$$

Hence f is differentiable and $f'(x) = g(x)$; we take for granted the fact that the indefinite integral of a continuous function has the integrand as derivative.

Later (p. 138) we shall prove a more general theorem in which no continuity is assumed for f_n'. There is some point in doing so, since a function f can have a derivative at every point with the derivative not integrable, either in the

Riemann or the Lebesgue sense. An example is furnished by the function defined by $f(x) = x^2 \sin(1/x^2)$ for $x \neq 0$, $f(0) = 0$, for which f' is unbounded and hence not Riemann integrable; it is not Lebesgue integrable either.

Exercise 17.1. If f is a function from R_1 to R_1, having derivatives of all orders, and the series

$$\cdots + \int_0^x dt \int_0^t f(u)\,du + \int_0^x f(t)\,dt + f(x) + f'(x) + \cdots$$

converges uniformly, what is its sum?[26]

Our various applications of uniform convergence have been precise formulations under various circumstances of the principle that we can take limits term by term in a uniformly convergent sequence. Here is another one.

Exercise 17.1a. Prove Tannery's theorem[26a]: if $f_n(k) \to L_n$ as $k \to \infty$ for each n, and $|f_n(k)| \leqslant M_n$ for all k, where ΣM_n converges, then if $p = p(k) \to \infty$ when $k \to \infty$,

$$\lim_{k \to \infty} \{ f_1(k) + f_2(k) + \cdots + f_p(k) \} = \sum_{n=1}^{\infty} L_n.$$

Exercise 17.1b. Application: show that

$$\lim_{k \to \infty} (1 + x/k)^k = \sum_{n=0}^{\infty} x^n/n!.$$

There are fairly simple situations in which the theorem of p. 108 on term by term integration of a uniformly convergent sequence is inadequate. For example, we have $1 - x + x^2 - \cdots + (-x)^n = [1 - (-x)^{n+1}]/(1 + x)$, and hence $1 - x + x^2 - \cdots = 1/(1 + x)$ if $|x| < 1$. Formally, then,

$$\log 2 = \int_0^1 \frac{dx}{1 + x} = \int_0^1 dx - \int_0^1 x \, dx + \int_0^1 x^2 \, dx - \cdots$$

$$= 1 - \tfrac{1}{2} + \tfrac{1}{3} - \tfrac{1}{4} + \cdots.$$

We cannot justify this calculation by the theorem on uniform convergence, since the sequence $\{f_n\}$ with $f_n(x) = [1 - (-x)^n]/(1 + x)$ does not converge uniformly on $[0, 1]$ (since it does not converge at 1), and does not converge uniformly even on $[0, 1)$. Indeed, if it did, the supremum on $[0, 1)$ of $|x^n|/(1 + x)$ would approach zero as $n \to \infty$; but since $|x|^n/(1 + x) \geqslant \tfrac{1}{2}|x|^n$, the supremum in question is at least $\tfrac{1}{2}$ and so cannot approach zero. In this case it is easy to verify the result of the formal calculation without appealing to the theorem on bounded convergence (which applies since $|[1 - (-x)^n]/(1 + x)| \leqslant 2$). We have

$$1 - \frac{1}{2} + \frac{1}{3} - \cdots + \frac{(-1)^{n-1}}{n} = \int_0^1 \frac{1 - (-x)^n}{1 + x} \, dx$$

$$= \int_0^1 \frac{dx}{1 + x} - \int_0^1 \frac{(-x)^n}{1 + x} \, dx,$$

$$\left| 1 - \frac{1}{2} + \frac{1}{3} - \cdots + \frac{(-1)^{n-1}}{n} - \log 2 \right| \leqslant \left| \int_0^1 \frac{(-x)^n}{1 + x} \, dx \right|$$

$$\leqslant \int_0^1 \frac{x^n}{1 + x} \, dx \leqslant \int_0^1 x^n \, dx = \frac{1}{n + 1} \to 0.$$

This calculation shows simultaneously that the series $1 - \tfrac{1}{2} + \tfrac{1}{3} - \cdots$ converges and that its sum is $\log 2$.

The reader might reasonably conjecture at this point that if formal integration of a convergent sequence yields a convergent result, it is necessarily the correct result. This conjecture is not true, although since it happens to be true for power series it is hard to give a counterexample that does not appear somewhat artificial. However, if f_n is the function defined by $f_n(x) = n$ for $0 < x < 1/n$, and $f_n(x) = 0$ elsewhere, we have $f_n(x) \to 0$ for each x, while

$$\int_0^1 f_n(x) \, dx = 1,$$

so that

$$\lim \int_0^1 f_n(x) \, dx \quad \text{and} \quad \int_0^1 \lim f_n(x) \, dx$$

both exist and are different.

Exercise 17.2. Find an example of the same phenomenon with continuous functions f_n.

Similarly, it is not very difficult (although we shall not do it) to show that

$$\frac{\pi - x}{2} = \sum_{n=1}^{\infty} \frac{\sin nx}{n}, \qquad 0 < x < \pi.$$

Formal integration yields

$$\tfrac{1}{4} \pi^2 = \int_0^{\pi} \frac{\pi - x}{2} \, dx = \sum_{n=1}^{\infty} \frac{1}{n} \int_0^{\pi} \sin nx \, dx$$

$$= \sum_{n=1}^{\infty} \frac{1 - \cos n\pi}{n^2} = \sum_{n=1}^{\infty} \frac{2}{(2n+1)^2},$$

so that we have a simple way of summing the numerical series

$$\frac{1}{1^2} + \frac{1}{3^2} + \frac{1}{5^2} + \cdots.$$

Since the original series can be shown not to be uniformly convergent, the justification of our evaluation of the numerical series eludes the elementary theory. On the other hand, the series $\Sigma n^{-1}\sin nx$ can be shown to be boundedly convergent, so that the theorem about integrating a boundedly convergent sequence would be enough.

18. Pointwise limits of continuous functions.[27] We consider functions from an interval in R_1 to R_1. Although a *pointwise limit of continuous functions* need not be continuous, it is nevertheless true, as we now show, that it cannot be extremely discontinuous: *its points of continuity must at least form an everywhere dense set.* (Hence, for example, the everywhere discontinuous function mentioned on p. 78, which was obtained from continuous functions by two successive limiting processes, cannot be obtained by one limiting process involving only continuous functions.)

We begin by observing that if a function f is discontinuous at a point x, the images of arbitrarily small neighborhoods of x do not have arbitrarily small diameters. That is, there is an integer n such that the diameter of the image of each neighborhood of x is at least $1/n$. (The images of larger neighborhoods have, if anything, larger diameters, so we can say "every neighborhood" instead of "small neighborhoods.") Now let f be discontinuous at every point of an interval, and let E_n be the set of points x in this interval for which the diameter of the image of every neighborhood of x is at least $1/n$. As we have just seen, every x belongs to some E_n. Moreover, E_n is a closed set. For, if y is a limit point of E_n, every neighborhood of y contains some point x of E_n, and hence contains a neighborhood of x, and so the diameter of the image of every neighborhood of y is also at least $1/n$.

We now use Baire's theorem, which tells us that some E_n is dense in some subinterval J. Since E_n is closed, E_n contains J. That is, we have found an interval J with the property that the image of every subinterval of J has diameter at least $1/n$. The existence of such an interval J is, then, a consequence of the property of being discontinuous at every point of some interval.

We shall now show that a pointwise limit of continuous functions cannot have such intervals J, and hence cannot be discontinuous at every point of any interval.

The range of f, being some subset of R_1, can be covered by a countable number of intervals $I_n = (a_n, b_n)$, each of length less than $1/n$. Let us look at the inverse images H_n of these I_n. The sets H_n collectively cover the interval J, but none of them can contain a subinterval of J, since the images of subintervals of J all have diameters greater than $1/n$. On the other hand, by Baire's theorem one of the H_n is dense in a subinterval of J. If we knew that the H_n were closed we should have a contradiction, since a closed set that is dense in an interval contains that interval.

Even when f is a pointwise limit of continuous functions, there is no reason to suppose that the H_n are closed. However, we can show that each H_n is a countable union of closed sets, and this property will do just as well. For, if the H_n have this property, then, by still another application of Baire's theorem, one of the closed sets is dense in a subinterval of J, and so contains that subinterval. Since the closed sets are subsets of H_n, it follows that H_n will also contain a subinterval of J.

We have thus reduced the proof of our theorem to showing that, if f is a pointwise limit of continuous functions f_k, the sets H_n are countable unions of closed sets. We recall that H_n is the inverse image under f of the interval (a_n, b_n); that is, H_n is the set of points x such that $a_n < f(x) < b_n$. Consider an x in H_n. Then if j is large enough we have $a_n + 1/(2j) \leqslant f(x) \leqslant b_n - 1/(2j)$, since $a_n < f(x) < b_n$. Since $f_k(x) \to f(x)$, we then have $a_n + 1/j \leqslant f_k(x) \leqslant b_n - 1/j$ for all sufficiently large k. Let $E_{k,j}$ be the set on which this last inequality holds, and let $F_{m,j}$ be the intersection of $E_{m,j}$, $E_{m+1,j}$, $E_{m+2,j}$, and so on. The sets $E_{k,j}$ are closed because f_k is continuous (they are inverse images of closed intervals), and the sets $F_{m,j}$ are closed as intersections of closed sets. We have just seen that if x is in H_n, then x is in some $F_{m,j}$. That is, H_n is a subset of the union of all the $F_{m,j}$. On the other hand, if x is in some $F_{m,j}$ we have $a_n + 1/j \leqslant f_k(x) \leqslant b_n - 1/j$ for some j and all sufficiently large k; since $f_k(x) \to f(x)$, this implies that $a_n + 1/j \leqslant f(x) \leqslant b_n - 1/j$ and so $x \in H_n$. Thus H_n is precisely the union of all the $F_{m,j}$ for all positive integers m and j, so we have succeeded in representing H_n as a countable union of closed sets. This is what was required to complete the proof.

Only slight modifications of the proof show that a limit of continuous functions has points of continuity everywhere dense in every nonempty perfect set. In this form the condition can be shown to be reversible: a function whose points of continuity are everywhere dense in every nonempty perfect set can be represented as a limit of continuous functions.

Baire described discontinuous limits of continuous functions as of class 1; limits of functions of class 1, not themselves of class 1, as of class 2; and so on. There are functions that do not belong to any Baire class.

An interesting example of a function of Baire class 1 is any discontinuous function on R_2 that is continuous on each line parallel to a coordinate axis.[27a]

Exercise 18.1.   Construct an example of such a function.

It is interesting that although we asserted only that a pointwise limit of continuous functions has an everywhere dense set of points of continuity, more than this is true: its set of points of discontinuity must form only a set of the first category. This property is really independent of the fact that the function in question is a pointwise limit of continuous functions. In fact, we shall show that a real-valued function f whose domain is an interval in R_1, if continuous at the points of an everywhere dense set, is continuous except on a set of first category.

Consider the sets E_n of points x such that there exists a sequence $\{y_k\}$ whose limit is x and which has the property that $|f(y_k) - f(x)| > 1/n$. Every point of discontinuity is in some E_n; that is, the set of points of discontinuity is in the union of the E_n. Since there are countably many E_n, our theorem is proved if we show that each E_n is nowhere dense. If some E_n failed to be nowhere dense, some point x where f is continuous would be a limit point of that E_n. If we choose a positive δ such that $|y - x| < \delta$ implies that $|f(y) - f(x)| < 1/(2n)$, and then choose a point w of E_n so that $|w - x| < \frac{1}{2}\delta$, and let y_k be the points occurring in the definition of E_n, we have

$$|f(y_k) - f(w)| \leqslant |f(y_k) - f(x)| + |f(x) - f(w)| < 1/n$$

for sufficiently large k, contradicting the definition of E_n.

19. Approximations to continuous functions. We have seen (§10) that even though a function from R_1 to R_1 is continuous, its graph may be rather irregular, at least to the extent of having oscillations in every interval. On the other hand, there is always a quite smooth function whose graph is very close to the graph of a given continuous function. More precisely, *if the domain of a continuous function is a compact interval, we can find, as close as we please to the function, either a step function, or a continuous polygonal function, or a polynomial.* A step function has a graph made up of a finite number of horizontal line segments; a polygonal function has a graph made up of a finite number of line segments of any orientation (not vertical). Here "as close as we please" is to be interpreted in the metric of the space B. In other words, if f is the continuous function, and ϵ is a given positive number, there are a step function f_1, a continuous polygonal function f_2, and a polynomial f_3 such that $|f(x) - f_k(x)| < \epsilon$ ($k = 1, 2, 3$) for all x in the interval in question.

The property that makes such approximations possible is called uniform continuity. A function f is continuous at x if, given a positive ϵ, there is a positive δ such that $|f(x) - f(y)| < \epsilon$ if $|x - y| < \delta$. Here we shall in general have to take smaller and smaller δ's as we consider different x's. If it is always possible (for a given f) to find a δ that will work simultaneously for *all* x in a given set, the function f is said to be *uniformly continuous* on that set. We now show that *a continuous function is uniformly continuous on any compact set in its domain.* Indeed, we shall establish the same theorem in a more general setting: a function whose domain and range are in metric spaces is uniformly continuous on any compact subset S of its domain.

To prove this theorem, let ϵ be a positive number and attach to x first a neighborhood N such that $d(f(x), f(y)) < \epsilon$ for all y in N, and then the neighborhood M of center x and half the radius of N. The neighborhoods M cover S, and since S is compact, some finite number of them still cover S. Let these covering neighborhoods be M_1, $M_2, \ldots, M_n$. Let δ be the smallest radius of any M_k. Now let x and y be any two points of S such that $d(x, y) < \delta$. Since x is in some M_k, there is a point z that is the center of an M_k in which x lies. Then $d(y, z) \leqslant d(x, z) + \delta$.

Since δ is the smallest radius of an M_k, this inequality shows that y is in the M_k whose center is z. Hence by the way in which the M_k were constructed,

$$d(f(x), f(z)) \leqslant \epsilon$$

and

$$d(f(y), f(z)) \leqslant \epsilon,$$

so that by the triangle inequality

$$d(f(x), f(y)) \leqslant 2\epsilon.$$

Exercise 19.1. Can a function f be simultaneously uniformly continuous and unbounded on $0 < x \leqslant 1$?

Exercise 19.2. If f is continuous for $x \geqslant 0$ and $f(x) \to L$ (finite) as $x \to +\infty$, must f be uniformly continuous for $x \geqslant 0$?

Exercise 19.3. Show that if f is continuous for $x \geqslant 0$ and $f(x) - x \to L$ (finite) as $x \to +\infty$, then f is uniformly continuous on $x \geqslant 0$.

Exercise 19.4. If f is uniformly continuous for $x \geqslant 0$, must $f(x)$ approach a finite or infinite limit as $x \to +\infty$?

We now construct the three approximating functions whose existence was asserted above.

Let S be a compact interval $[a, b]$ in R_1. We already have S covered by a finite number of intervals (the M_k of the preceding proof), on each of which the given function f varies by less than ϵ, and we want to construct a step function f_1 such that $f_1(x)$ differs from $f(x)$, for each x, by less than ϵ. The M_k (being open intervals) necessarily overlap. If they didn't overlap, we could take $f_1(x)$ to equal $f(x_k)$ where x_k is the midpoint of M_k, so we need a systematic way to discard the overlapping pieces. A straightforward but rather inelegant method is as follows. Some M_k (perhaps more than one) covers the point a. Discard all but the one of these that reaches farthest toward b, and call it N_1. If N_1 contains b, stop. Otherwise there is an M_k that overlaps N_1 and extends farthest to the right. Discard the overlap (except for its left-hand endpoint), and call the reduced (half-open) interval N_2. If N_2 contains b, stop. Otherwise continue in the same way. The process terminates because there were only finitely many M_k to begin with, and we have finitely many N_j that cover S. Since each N_j is a subset of some M_k, $f(x)$ varies by less than ϵ on each N_j. Consequently, if we define $f_1(x) = f(x_j)$ on N_j, where x_j is (for example) the center of the M_k from which N_j was obtained, then $|f_1(x) - f(x)| < \epsilon$ on S.

To construct the polygonal approximation f_2, shorten each step of f_1 by a small amount at each end, and then join the right-hand endpoint of each reduced step to the left-hand endpoint of the next reduced step by a line segment.

To construct a polynomial approximation f_3 is somewhat harder.[28] One procedure is as follows. We may suppose, merely in order to simplify the formulas, that the domain of our given continuous function is an interval $[h, 1 - h]$, where $0 < h < 1$. We can extend our function in an obvious way so that the extended function f is continuous on R_1 and is zero outside $(\frac{1}{2}h, 1 - \frac{1}{2}h)$. Consider now the function I defined by

$$I(x) = c_n \int_0^1 f(t) \left[1 - (x - t)^2 \right]^n dt,$$

where

$$1/c_n = \int_{-1}^1 (1 - t^2)^n dt.$$

Evidently I is a polynomial of degree $2n$. The factor in brackets in the integrand has a peak at $t = x$ and (when n is large) is small when t is not near x, so that it may seem plausible that $I(x)$ should be close to $f(x)$. We now show that this is really the case.

We can write

$$I(x) = c_n \int_{x-1}^x f(x - s)(1 - s^2)^n ds,$$

and since $f(t) = 0$ when $t < 0$ or $t > 1$, this is the same as

$$I(x) = c_n \int_{-1}^1 f(x - s)(1 - s^2)^n ds.$$

Next, we can write, because of the way c_n was defined,

$$I(x) - f(x) = c_n \int_{-1}^1 \left[f(x - s) - f(x) \right] (1 - s^2)^n ds.$$

We now break the integral up into three parts,

$$I_1 = c_n \int_{-1}^{-\delta}, \quad I_2 = c_n \int_{-\delta}^{\delta}, \quad I_3 = c_n \int_{\delta}^{1},$$

where $0 < \delta < 1$ and δ is to be chosen in a moment.

At this point we use the uniform continuity of f. If ϵ is an arbitrary positive number we can find δ small enough so that $|f(x - s) - f(x)| < \epsilon/3$ if $|s| < \delta$, where the inequality holds for all x with the same δ. We use this inequality to estimate I_2:

$$|I_2| \leqslant \frac{1}{3} \epsilon c_n \int_{-\delta}^{\delta} (1 - s^2)^n \, ds$$

$$< \frac{1}{3} \epsilon c_n \int_{-1}^{1} (1 - s^2)^n \, ds = \epsilon/3.$$

Next, since f is continuous on a compact set, it is bounded, say $|f(x)| \leqslant M$. In I_3 we have $(1 - s^2)^n \leqslant (1 - \delta^2)^n$, while

$$1/c_n = \int_{-1}^{1} (1 - t^2)^n \, dt \geqslant \int_{0}^{\delta/2} (1 - t^2)^n \, dt \geqslant \tfrac{1}{2}\delta\left(1 - \tfrac{1}{4}\delta^2\right)^n.$$

Hence we have

$$|I_3| \leqslant 2c_n M \int_{\delta}^{1} (1 - s^2)^n \, ds$$

$$\leqslant 4M\delta^{-1}(1 - \delta)(1 - \delta^2)^n \left(1 - \tfrac{1}{4}\delta^2\right)^{-n}.$$

Since $(1 - \delta^2)/(1 - \tfrac{1}{4}\delta^2) < 1$, this estimate shows that $I_3 \to 0$ as $n \to \infty$. Exactly similar reasoning applies to I_1. By taking n large enough we then have $|I_1|$ and $|I_3|$ less than $\epsilon/3$. Hence

$$|I(x) - f(x)| \leqslant |I_1| + |I_2| + |I_3| \leqslant \epsilon,$$

provided that n is large enough. Hence the polynomial I furnishes the approximation f_3 if n is large enough.

The possibility of approximating a continuous function by polynomials has many applications. It was used on p. 69 in the proof of the existence of continuous nowhere differentiable functions. Here is another application.

Let f be continuous on the interval $[a, b]$. The quantities $\int_a^b f(x)x^n \, dx$ $(n = 0, 1, 2, \ldots)$ are called the moments of f. We shall show that a continuous function with a compact domain in R_1 is determined by its moments; that is, *two continuous functions with the same sequence of moments are identical*. (We do not say anything about how a continuous function can actually be calculated from its moments.) It is an equivalent statement that a continuous function all of whose moments are zero must vanish identically, and this we now prove. Suppose that all the moments of f are zero. Without loss of generality, we suppose $a = 0$, $b = 1$. If f is not identically zero, it is positive (or negative) in some interval, and we can construct a continuous function g that is zero outside this interval and makes $\int_0^1 fg \, dx = 2h > 0$. (See the figure.) Construct a polynomial P such that $|g(x) - P(x)| < h/\max|f(x)|$. Then

$$\int_0^1 fP \, dx = \int_0^1 fg \, dx - \int_0^1 f \cdot (g - P) \, dx$$

$$\geqslant 2h - \max|f(x)| \cdot \max|g(x) - P(x)| > h.$$

But $\int_0^1 fP \, dx = 0$ since all the moments of f vanish. The contradiction can be avoided only by having f identically zero.

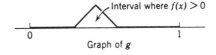

Interval where $f(x) > 0$

0   1

Graph of **g**

As a corollary of the theorem about moments, we see that the set of all continuous functions can be put into one-to-one correspondence with a class of sequences of real numbers, since different continuous functions have different sequences of moments. Since the number of sequences of real numbers is the same as the number of real numbers (Exercise 3.10), it follows that there are just as many continuous functions as there are real numbers. (A more direct way of seeing this is to observe that a continuous function is determined by its values at the rational points, that is, by a sequence of real numbers.)

The property of uniform continuity is also useful in justifying interchanges of limiting operations in calculus. For example, let f be a continuous function with domain in R_2 and suppose that $f(x, b) = L$ (independent of x) for all x in some interval; that is, the function is constant along the horizontal line segment $y = b$. It is then not necessarily true that as $y \to b$ the partial derivative $\partial f/\partial x$ (evaluated at (x, y)) approaches zero for each x. (Example: $f(x, y) = y \sin(1/(xy))$, $f(x, 0) = 0$, $b = 0$, $L = 0$; the value of $\partial f/\partial x$ at (x, y) is $-x^{-2}\cos(1/(xy))$, which does not approach a limit.) That is, we cannot necessarily say that

$$\lim_{y \to b} \lim_{h \to 0} \frac{f(x + h, y) - f(x, y)}{h} = \lim_{h \to 0} \lim_{y \to b} \frac{f(x + h, y) - f(x, y)}{h}$$

$$= \lim_{h \to 0} \frac{f(x + h, b) - f(x, b)}{h} \,.$$

However, we can make this statement in any neighborhood of a point (a, b) in which $\partial f/\partial x$ is a continuous function (that is, continuous as a function with domain a neighborhood in R_2, not merely "continuous in x for each y and continuous in y for each x"; the latter means only continuity (in R_1) of the restrictions of f to lines parallel to the coordinate axes). The proof depends on the uniform continuity of $\partial f/\partial x$. First, $\partial f/\partial x$ has the value 0 at any (x, b) since $f(x, b) = L$. Next, by the law of the mean we have

$$\frac{f(x + h, y) - f(x, y)}{h} = \frac{\partial f}{\partial x}(x + h', y),$$

with h' between 0 and h. Finally, $\partial f / \partial x$, evaluated at $(x + h', y)$, differs from $\partial f / \partial x$, evaluated at (x, b) (where it is 0), by arbitrarily little provided that h (and so h') and $|y - b|$ are sufficiently close to 0.

20. Linear functions.   A function f whose domain is R_1 is said to be *linear* if $f(x) + f(y) = f(x + y)$ for all x and y. (This is a more special use of the term linear than is often made: the function f defined by $f(x) = ax + b$ is not linear in our sense when $b \neq 0$.) Clearly if $f(x) = ax$ then f is linear, and we might expect that all linear functions would be of this form; but not all of them are. To exhibit one that is not, we should have to appeal to one of the more abstruse properties of the real number system, which depends on ideas that have not been introduced in this book.[29] What is also not immediately obvious, but will be established shortly, is that a discontinuous linear function is necessarily wildly discontinuous: it is, for example, unbounded in every interval, and indeed its graph must be everywhere dense in R_2. It is only to be expected that no very simple construction will produce a function of this kind.

Let us consider a linear function f. For every x we have $f(2x) = f(x + x) = 2f(x)$, and so by induction $f(nx) = nf(x)$ for every positive integer n. Since $f(x) = f(x + 0) = f(x) + f(0)$, we have $f(0) = 0$. Since $0 = f(0) = f(x - x) = f(x) + f(-x)$, we have $f(-x) = -f(x)$. Therefore $f(nx) = nf(x)$ for every integer, positive or negative. Replacing x by x/n we obtain $f(x) = nf(x/n)$, or $f(x/n) = n^{-1}f(x)$. Now replace x by mx, and we have $f(mx/n) = n^{-1}f(mx) = (m/n)f(x)$. In other words, $f(rx) = rf(x)$ for every rational number r. In particular (put $x = 1$), $f(r) = rf(1)$ for every rational number r. It follows that if f is continuous for all x then $f(x) = xf(1)$ for all x.

We can easily strengthen this result by showing that

$f(x) = xf(1)$ for all x provided merely that f is continuous at some point c. For, if $a < c < b$, we have $f(c + \delta) - f(c) \to 0$ as $\delta \to 0$; but $f(c + \delta) - f(c) = f(\delta)$, so $f(\delta) \to 0$ as $\delta \to 0$. That is, f is continuous at 0. Now if x is any real number, $f(x + \delta) - f(x) = f(\delta) \to 0$ as $\delta \to 0$, so f is continuous at x. Thus f is continuous throughout R_1, and we know that this implies that $f(x) = xf(1)$.

We can go still further in weakening the hypothesis and nevertheless being able to prove that a linear function is continuous. Suppose that f is merely bounded on some interval, or even on some set E having the following property: the set of all distances $|x - y|$ between points x and y of E contains a neighborhood of 0. That is, there is a positive δ such that if $|t| < \delta$ then there are points x and y in E such that $x - y = t$. Then we can still conclude[30] that f is continuous if it is linear, and so a linear f that is bounded on a set of the kind just described must be of the form $f(x) = cx$.

Let $|f(x)| \leqslant M$ on E. For numbers t that are distances between points of E, $|f(t)| = |f(x - y)| = |f(x) - f(y)| \leqslant 2M$. Therefore if $|u| < \delta/n$, we have $|f(u)| = n^{-1}|f(nu)| \leqslant 2M/n$. Now let s be any real number, and let r be a rational number such that $|r - s| < \delta/n$. Then we have

$$|f(s) - sf(1)| = |f(s - r) + (r - s)f(1)|$$

$$\leqslant 2M/n + \delta|f(1)|/n.$$

Since n can be as large as we please, we have $f(s) - sf(1) = 0$.

There are many sets E, other than intervals, having the property used in this proof. They include the so-called sets of positive measure (for which we must refer to works on Lebesgue integration), and some sets of measure 0. For example, the Cantor set (p. 37) has the property. The proof of this fact can be given a very intuitive geometrical form.[31] Take the usual coordinate system in R_2 and construct Cantor sets on the intervals $[0, 1]$ of both the x and y axes, removing from the plane not only the middle thirds of intervals, but also all the points of the unit square that have one coordinate (at least) in a deleted interval, so that at each step we remove some cross-shaped sets. Consider a line with equation $y = x + c$, where $0 \leqslant c \leqslant 1$. At each step, the line meets at least one of the squares that is not deleted in this step; these squares are closed, and nested, so their intersection contains some point (x, y) with $y = x + c$, and both x and y are points of the Cantor set.

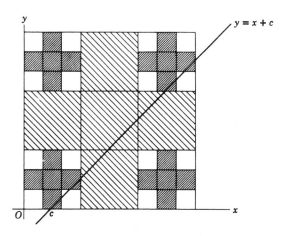

To show[32] that $f(x) = cx$ for a linear function f whose graph is not everywhere dense in R_2, we may appeal to an elementary fact from the theory of numbers. Let (x_1, x_2) and (y_1, y_2) be two pairs of real numbers that are not proportional; this means that $x_1 y_2 \neq x_2 y_1$, or in geometrical language that the points (x_1, x_2) and (y_1, y_2) of R_2 are not on the same straight line through the origin. Then if a and b are any two real numbers whatsoever, we can find rational multipliers r_1 and r_2 so that $r_1x_1 + r_2x_2$ is as close as we like to a and simultaneously $r_1y_1 + r_2y_2$ is as close as we like to b. To prove this, we solve the equations $ux_1 + vx_2 = a$ and $uy_1 + vy_2 = b$ (since their determinant is not zero), and then choose r_1 and r_2 close to u and v, respectively.

Now suppose that f is linear and not of the form $f(x) = ax$. The second hypothesis implies that we must be able to find points x_1, x_2 such that $f(x_1)/x_1 \neq f(x_2)/x_2$. Then for any point (a, b) in R_2 we can find rational numbers r_1 and r_2 such that $f(r_1x_1 + r_2x_2) = r_1 f(x_1) + r_2 f(x_2)$ differs arbitrarily little from b, and at the same time $r_1x_1 + r_2x_2$ differs arbitrarily little from a. Thus there is a point of the graph of f as close as we please to the point (a, b) of R_2.

Another proof that yields some additional insight can be given as follows. Suppose that f is linear but not of the form $f(x) = cx$. Since $f(r + t) = f(r) + f(t)$, and $f(r) = cr$ for rational r, it is enough to show that f takes, arbitrarily close to the origin, values arbitrarily close to every real number, or, what is the same thing, arbitrarily close to every rational number. Let A be a positive rational number and $\epsilon (< 1)$ a small positive number that we use to specify closeness. We know that f is unbounded in $(0, \epsilon)$; suppose, for definiteness, that f takes arbitrarily large positive values there. Then there is an integer n exceeding A/ϵ such that for some s in $(0, \epsilon)$ we have $n + 1 \geqslant f(s) \geqslant n$. Since $f(rx) = rf(x)$ for all rational r and all x, we have $f(As/n) = (A/n)f(s)$, so that $A + \epsilon > A(n+1)/n \geqslant f(As/n) \geqslant A$, and As/n is a point of the interval $(0, \epsilon)$. We have therefore constructed a point arbitrarily close to 0 where f takes a value arbitrarily close to A, if $A > 0$. If

$A < 0$, there is a point close to 0 where f takes a value close to $-A$, and since $f(-x) = -f(x)$, the same conclusion follows.

Here is an application in calculus of our theorems about linear functions.[33] *Suppose that the limit*

$$\lim_{R \to \infty} \frac{1}{2R} \int_{x-R}^{x+R} f(u)\, du$$

exists for every real x; *denote it by* $\phi(x)$. We shall now show that $\phi(x)$ *is necessarily of the form* $ax + b$. We have, first,

$$\int_{(x-h)-R}^{(x-h)+R} f(u)\, du + \int_{(x+h)-R}^{(x+h)+R} f(u)\, du$$

$$= \int_{x-(R+h)}^{x+(R+h)} f(u)\, du + \int_{x-(R-h)}^{x+(R-h)} f(u)\, du,$$

and so $\phi(x - h) + \phi(x + h) = 2\phi(x)$. If we replace $x - h$ and $x + h$ by x and y, this says that $\phi(x) + \phi(y) = 2\phi(\frac{1}{2}(x + y))$. Put $\phi(x) - \phi(0) = \psi(x)$; then we have

$$\psi(x) + \psi(y) = \phi(x) + \phi(y) - 2\phi(0)$$

$$= 2\phi\big(\tfrac{1}{2}(x + y)\big) - 2\phi(0) = 2\psi\big(\tfrac{1}{2}(x + y)\big)$$

$$(*)$$

This is true for every y, and so in particular for $y = 0$; putting $y = 0$, we find $\psi(0) = 0$ and $\psi(x) = 2\psi(x/2)$. In the last equation replace x by $x + y$; then it says that $\psi(x + y) = 2\psi(\frac{1}{2}(x + y))$. However, $(*)$ above says that $2\psi(\frac{1}{2}(x + y)) = \psi(x) + \psi(y)$. Therefore $\psi(x) + \psi(y) = \psi(x + y)$; that is, ψ is linear. Now ψ is a limit of continuous functions, and so must have points of continuity (§18). But we know that a linear function ψ that has a point of continuity has the form $\psi(x) = x\psi(1)$, which is to say that $\phi(x) = \phi(0) + x\psi(1)$, as asserted.

Exercise 20.1.[33a] Let $\phi(x)$ denote

$$\lim_{R \to \infty} \int_R^{x+R} f(u)\, du,$$

supposed to exist for every real x. Then $\phi(x) = ax$.

21. Derivatives.[33b] We consider only functions whose ranges are in R_1 and whose domains are intervals in R_1. Along with the derivative of a function f, which can be defined in the usual way, we shall consider some generalizations that have the advantage of applying to functions that are not necessarily differentiable in the usual sense. These are the four *Dini derivates*, for which we shall use the following notations and definitions:

$$f^+(x) = \limsup_{h \to 0+} \frac{f(x+h) - f(x)}{h} ,$$

$$f_+(x) = \liminf_{h \to 0+} \frac{f(x+h) - f(x)}{h} ,$$

$$f^-(x) = \limsup_{h \to 0-} \frac{f(x+h) - f(x)}{h} ,$$

$$f_-(x) = \liminf_{h \to 0-} \frac{f(x+h) - f(x)}{h} ;$$

the $+$ and $-$ refer to right and left, respectively, and their (upper or lower) positions refer to upper and lower limits. For each x, the four derivates exist, finite or infinite, for any function f at all.

It is common practice to use the phrase "the derivative of f" to mean, according to context, either the number $f'(x)$, that is, the derivative of f at the particular point x, or the function f' whose value at x is the number $f'(x)$. We shall use the same ambiguous terminology for the Dini derivates. If we are going to talk about them as functions, we have to extend our usual notion of function by considering functions whose values may include $+\infty$ or $-\infty$. We must be careful about such generalized functions; there are difficulties about forming sums or products, or (for example) about trying to differentiate them. It will be found that we do not in fact do anything ambiguous with derivates.

If $f^+(x) = f_+(x)$, we say that there is a *right-hand derivative* at x, and denote it by $f'_+(x)$; similarly for the left-hand derivative $f'_-(x)$. Finally, the ordinary derivative $f'(x)$ exists (finite or infinite) if and only if all four derivates are equal.

Even when $f^+(x)$ and $g^+(x)$ are both finite, we do not necessarily have $(f + g)^+(x) = f^+(x) + g^+(x)$; but if $f'(x)$ exists we do have $(f + g)^+(x) = f'(x) + g^+(x)$ (cf. §15).

Exercise 21.1. Show that, if $f'_+(x)$ exists and is finite, f is continuous on the right at x; and that, if $f'(x)$ exists and is finite, f is continuous at x.

Exercise 21.2. Show that f may be discontinuous at x when $f'(x)$ exists and is infinite.

On the other hand, we have already seen that a continuous function does not have to have a derivative anywhere (finite or infinite).

Exercise 21.3. Show that if $f'(a)$ exists (finite), we can write $f(x) - f(a) = (x - a)[f'(a) + \epsilon(x)]$, where $\lim_{x \to a} \epsilon(x) = 0$.

The "chain rule" for differentiation states that if $f'(a)$ exists (finite), if $g(b) = a$, and if $g'(b)$ exists (finite), then for the function ϕ such that $\phi(x) = f(g(x))$, the derivative

$\phi'(b)$ exists and is equal to $f'(a)g'(b)$. A fallacious proof proceeds as follows: as $h \to 0$,

$$\frac{\phi(x+h) - \phi(x)}{h} = \frac{f(g(x+h)) - f(g(x))}{g(x+h) - g(x)} \frac{g(x+h) - g(x)}{h}$$

$$\to f'(g(b)) g'(a).$$

Exercise 21.4. Find the fallacy; give a correct proof by using Exercise 21.3.

Another application of Exercise 21.3 is the following necessary condition for finite differentiability (which is also sufficient).[33c] If the domain of f contains an interval in R_1 and a is an interior point of this interval, then for every positive ϵ there is a neighborhood N of a that is so small that whenever t_1 and t_2 are on opposite sides of a and in N, and u_1 and u_2 are on opposite sides of a and in N, then

$$\left| \frac{f(t_1) - f(t_2)}{t_1 - t_2} - \frac{f(u_1) - f(u_2)}{u_1 - u_2} \right| < \epsilon.$$

It is enough to show that each of the fractions on the left can be made arbitrarily close to $f'(a)$. Now, for example,

$$\frac{f(t_1) - f(t_2)}{t_1 - t_2} = \frac{f(t_1) - f(a)}{t_1 - a} \frac{t_1 - a}{t_1 - t_2} - \frac{f(t_2) - f(a)}{t_2 - a} \frac{t_2 - a}{t_1 - t_2}$$

$$= (f'(a) + \epsilon_1) \frac{t_1 - a}{t_1 - t_2} - (f'(a) + \epsilon_2) \frac{t_2 - a}{t_1 - t_2}$$

$$= f'(a) + \epsilon_1 \frac{t_1 - a}{t_1 - t_2} - \epsilon_2 \frac{t_2 - a}{t_1 - t_2},$$

where ϵ_1 and $\epsilon_2 \to 0$ as t_1 and $t_2 \to a$. Since $|(t_1 - a)/(t_1 - t_2)| \leqslant 1$ and $|(t_2 - a)/(t_1 - t_2)| \leqslant 1$ we have what we need.

This last proposition can be used for verifying the non-differentiability of some continuous functions.[34] An example is as follows. Let $G(x)$ be the distance from the real number x to the nearest integer; the graph of G looks like this:

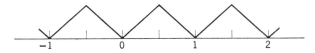

Let $H(x) = \sum_{n=0}^{\infty} 2^{-n} G_n(x)$, where $G_n(x) = G(2^n x)$. That H is continuous follows from the continuity of G and the uniform convergence of the series. Let a be any point and $(a - \delta, a + \delta)$ an interval containing a. Since the points $2^{-k} r$, where k runs through all positive integers and r through all integers, form an everywhere dense set, we can find points $x_1 = 2^{-k} r$, $x_2 = 2^{-k}(r + 1)$ in $(a - \delta, a + \delta)$ and on opposite sides of a. Let ξ be the midpoint of (x_1, x_2).

The graph of G has corners at the points $2^{-1} p$, where p runs through the integers; the graph of G_1 has corners at the points $2^{-2} p$, and generally the graph of G_n has corners at $2^{-n-1} p$. Thus the graphs of $G_0, G_1, \ldots, G_{k-1}$ have no corners between x_1 and x_2, and consequently the slope of each G_j (for $0 \leqslant j \leqslant k - 1$) is the same between x_1 and ξ as it is between x_1 and x_2. On the other hand, for $n > k$ we have $G_n(x_1) = G_n(x_2) = 0$. Thus

$$\frac{H(\xi) - H(x_1)}{\xi - x_1} - \frac{H(x_2) - H(x_1)}{x_2 - x_1}$$

reduces to

$$\frac{G_k(\xi) - G_k(x_1)}{\xi - x_1} - \frac{G_k(x_2) - G_k(x_1)}{x_2 - x_1} = \pm 1,$$

and a similar argument holds for

$$\frac{H(x_2) - H(\xi)}{x_2 - \xi} - \frac{H(x_2) - H(x_1)}{x_2 - x_1} .$$

Since either ξ and x_1, or ξ and x_2, are on opposite sides of a, our necessary condition for finite differentiability cannot be satisfied. This shows that H has no finite derivative at any point. (Our earlier construction of a nowhere differentiable function showed that a continuous function need not even have an infinite derivative at any point.)

Exercise 21.5.   Show that *if* $f'(x) > 0$, *then f is increasing at* x, in the sense that there is an interval $(x - h, x + h)$ such that if s and t are in the interval and $s < x < t$ then $f(s) < f(x) < f(t)$. More generally, if $f_+(x) > 0$, f is increasing on the right at x, in an obvious sense.

Exercise 21.5a.   A necessary and sufficient condition for $f'_+(x_0)$ to exist (finite or infinite) is that for every real number K, with at most one exception, $f(x) + Kx$ is monotonic on the right at x_0.[34a]

Exercise 21.5b.   The only continuous functions f for which $f(x) + Kx$ is monotonic on (a, b) for every real K, with at most one exception, are of the form $f(x) = px + q$ on (a, b).

Note, for comparison with Exercise 21.5b, that $f(x) = x^2 \sin(1/x)$, $f(0) = 0$, has $f(x) + Kx$ monotonic whenever $|K| > 3$.

We say that f has a *maximum* at x (an interior point of the domain of f) if there is a neighborhood N of x such that $f(y) \leqslant f(x)$ for all y in N; the maximum is *proper* if there is a neighborhood N' of x such that $f(y) < f(x)$ for y in N' and $y \neq x$.

Exercise 21.6.   Show that if f has a maximum at x, then $f^+(x) \leqslant 0$ and $f_-(x) \geqslant 0$.

In particular, *if f has a maximum at* x *and* $f'(x)$ *exists, we must have* $f'(x) = 0$. There are of course similar results for minima.

There are only a countable number of proper maxima of any function.[35] To see this, let us assign to a proper maximum of f, occurring, say, at x, an interval (r_1, r_2) with rational endpoints, such that r_1 and r_2 are on opposite sides of x and $f(y) < f(x)$ provided that $y \neq x$ and $r_1 < y < r_2$. This interval cannot also be assigned to some other proper maximum, occurring at z, since it would contain both z and x and we should have to have both $f(z) < f(x)$ and $f(x) < f(z)$. Consequently, different proper maxima have different rational intervals assigned to them. Since there are only a countable number of intervals with rational endpoints there are at most a countable number of proper maxima.

There can, however, be an uncountable number of improper maxima for a continuous function; for example, a constant function has improper maxima at all points. It can be shown that the values of any function at the points where its derivative is zero (or even where one of its Dini derivates is zero) form a set of measure zero.[36] Taken in conjunction with the general properties of derivates that will be mentioned on p. 147, this fact shows that *the ordinates of all maxima, proper or not, form a set of measure zero*.

We have observed that when a derivative exists at a maximum or minimum inside the domain of the function (supposed here to be an interval), the derivative must be equal to 0 at that point; hence if we want to show that a derivative takes the value 0 we frequently proceed by showing that the function from which it was derived has a maximum or minimum, not at an endpoint of its domain. If we want to show that f' assumes some other value c, we consider $g(x) = f(x) - cx$ and look for its maxima and minima. Any hypothesis that forces g to have a maximum or minimum in an interval (a, b) then guarantees that c is in the range of f'. Two hypotheses that do this for continuous functions f are (A) that $f'(a) > 0$ and $f'(b) < 0$ (since then $g(x) > g(a)$ in a right-hand neighborhood of a, and so the largest value of g between a and b is not attained at a; similarly at b); or (B) that $g(a) = g(b)$ (since then either g is constant or has a proper maximum or minimum between a and b). Hypothesis (A) leads to the observation that derivatives have the intermediate value property:[36a] *if a derivative takes two values, it takes every value between them.*

Exercise 21.7. Show that this follows from the remark jus made, that if f' takes a positive value and a negative value, it takes the value 0.

Exercise 21.8. Let f be a periodic differentiable function; let a be a given positive number; then there is a point x such that the tangent at x meets the graph again at a point a units farther along the x-axis (i.e., $f(x + a) - f(x) = af'(x)$).

Hypothesis (B) leads to the *mean-value theorem* (also known as the law of the mean), which states that every difference quotient $[f(x) - f(y)]/(x - y)$ of a differentiable function f is in the range of f' (the usual formulation is —superficially—different). The proof is suggested by the diagram[36b]:

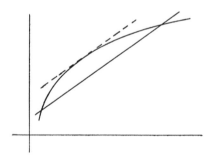

$$g(x) = f(x) - \frac{x - a}{b - a}\left[\, f(b) - f(a)\,\right]$$

takes the same value $f(a)$ at b and at a, so it has a maximum at some point c between a and b; at this point, $g' = 0$ and $f'(c) = [f(b) - f(a)]/(b - a)$.

A less conventional procedure[36c] is to start from $g(a) = g(b) = 0$ and infer from the universal chord theorem (p. 92) that there are intervals (x_n, y_n) in (a, b), each half as long as its predecessor, with $g(x_n) = g(y_n)$. These intervals are nested and hence converge to a point c, which will be in the open interval (a, b) if we pick the first two intervals to avoid a and b. Since we have a sequence of horizontal chords of g whose endpoints approach c, the tangent at c (assumed to exist) must be horizontal, i.e., $g'(c) = 0$.

One should not overemphasize the existence of the point c, whose location is usually unknown; what is generally wanted in practice is that $[f(b) - f(a)]/(b - a)$, the difference quotient, is between sup f' and inf f'; this is actually an equivalent property because f' has the intermediate value property.[36d] Another way of stating all this is to say that the range of f' is an interval which contains the range of the difference quotients $[f(x) - f(y)]/(x - y)$. The range of the difference quotients, however, does not necessarily contain the range of f'; in other words, the converse of the mean-value theorem may fail. An example is given by $f(x) = x^3$, where f' takes the value 0, whereas no difference quotient is 0. However, the least upper and greatest lower bounds of the set of values of difference quotients are the same as the corresponding bounds of f'. What we have to show is that (for example) it is impossible to have

$$\sigma = \sup_{(x,y)} \frac{f(x) - f(y)}{x - y} < S = \sup_t f'(t).$$

This would mean that there is a point t_0 such that $f'(t_0)$ exceeds all possible values of $[f(x) - f(y)]/(x - y)$ by nearly $S - \sigma$. Since $f'(t_0)$ is a limit of difference quotients, there is then a difference quotient that exceeds $f'(t_0) - \frac{1}{2}(S - \sigma)$, and we have a contradiction.

In the mean-value theorem we supposed that f is continuous in the closed interval $[a, b]$. We can, as a matter of fact, drop the continuity of f at the endpoints provided that we require continuity on the right at a and on the left at b in the case where the limits $f(a +)$ and $f(b -)$ exist, and otherwise require nothing at all at the endpoints. However, the greater generality so obtained is illusory, since if $f(a +)$ does not exist and f' is finite near a, then f' assumes every finite value in every right-hand neighborhood of a, so that $(b - a)f'(c)$ can have any finite value we please.[37] For, if k is any number, $f(x) - kx$ does not have a right-hand limit at a, and so cannot be monotonic in a right-hand neighborhood of a. It therefore has maxima and minima in every right-hand neighborhood of a, and its derivative is zero at such points x; then $f'(x) = k$.

As an application of the mean-value theorem we now prove a theorem on the termwise differentiation of a sequence of functions. The elementary theorem that we proved in §17 demands integrability of the derivatives, and its proof uses the theorem on the integration of a uniformly convergent sequence of functions; but it is possible to prove a more general theorem without using any integration at all. This is: *let the functions f_n have (finite) derivatives f_n' in an interval I; let the sequence $\{f_n(a)\}$ converge for some a in I, and let $\{f_n'\}$ converge uniformly, say to g. Then f_n converges to a limit f, uniformly on I if I is compact, otherwise uniformly on each compact subset of I; and $f'(x) = g(x)$ for all x in I.*

To prove this, first apply the mean-value theorem to $f_n - f_m$:

$$f_n(x) - f_m(x) - \left[f_n(a) - f_m(a) \right] = (x - a)\left[f_n'(c) - f_m'(c) \right],$$

where c is between x and a (and, of course, may depend on m and n). The uniform convergence of $\{f'_n\}$ and the convergence of $\{f_n(a)\}$ thus make $\{f_n\}$ converge uniformly as long as $|x - a|$ is bounded. Let f be the limit of f_n and let ϵ be an arbitrary positive number. We have

$$|f_n(x) - f_m(x) - [f_n(a) - f_m(a)]| \leqslant (x - a)\epsilon$$

if n and m exceed some integer n_0. Letting $m \to \infty$, we see that

$$|f_n(x) - f(x) - [f_n(a) - f(a)]| \leqslant (x - a)\epsilon, \quad n > n_0.$$

That is,

$$\left| \frac{f_n(x) - f_n(a)}{x - a} - \frac{f(x) - f(a)}{x - a} \right| \leqslant \epsilon, \quad n > n_0.$$

Also, we have, by the hypothesis of uniform convergence, $|f'_n(a) - g(a)| \leqslant \epsilon$ if $n > n_1$. Now fix an n that exceeds both n_0 and n_1. Then if $|x - a|$ is small enough,

$$\left| \frac{f_n(x) - f_n(a)}{x - a} - f'_n(a) \right| < \epsilon$$

and so

$$\left| \frac{f(x) - f(a)}{x - a} - f'_n(a) \right| < 2\epsilon$$

if $|x - a|$ is small enough. But $|f'_n(a) - g(a)| \leqslant \epsilon$, so

$$\left| \frac{f(x) - f(a)}{x - a} - g(a) \right| < 3\epsilon.$$

This inequality shows that $f'(a)$ exists and is equal to $g(a)$. Since we now know that $\{f_n\}$ converges uniformly, we can take a to be any point whatever in I, and the theorem follows.

Another application of the mean-value theorem yields the following theorem,[38] which has a transparent geometrical interpretation. *Suppose that f is differentiable in $[a, b]$ and that $f'(a) = f'(b)$; then there is a point c in (a, b) such that*

$$\frac{f(c) - f(a)}{c - a} = f'(c).$$

This says that if the graph of f has the same slope at a and at b, there must be a point c at which the tangent passes through the initial point $(a, f(a))$; a sketch will make this geometrically plausible.

In proving this, we may suppose that $f'(a) = f'(b) = 0$, since otherwise we should consider the function defined by $f(x) - xf'(a)$. Consider the function g defined by

$$g(x) = \frac{f(x) - f(a)}{x - a}, \quad a < x \le b; \; g(a) = 0.$$

The function g is continuous in $[a, b]$ (care is required in verifying this at the point a), and differentiable in $(a, b]$. We have $g'(b) = -g(b)/(b - a)$. If $g(b) > 0$, then we have $g'(b) < 0$ and hence g decreasing at b (Exercise 21.5), while $g(a) = 0$, so that g attains its maximum at a point c between a and b where $g'(c) = 0$. A similar argument applies if $g(b) < 0$. If $g(b) = 0$, we have $g(a) = g(b) = 0$ and again $g'(c) = 0$ for some intermediate c. Since

$$g'(c) = \frac{f'(c)}{c - a} - \frac{f(c) - f(a)}{(c - a)^2},$$

our conclusion follows.

Still another application of the mean-value theorem serves to justify the intuitive idea that derivatives tend to behave worse than the functions from which they are derived.[39]

Exercise 21.9. If $f(x) \geqslant 0$, $f(0 +) = 0$, f is differentiable in $(0, 1]$, $h(x) \geqslant 0$, and $\int h(x)\,dx$ diverges at 0, then $h(f(x))f'(x)$ is unbounded as $x \to 0$; for example, $f'(x)/f(x)$ is unbounded and so is $f'(x)/\{f(x)\log f(x)\}$ (provided $f(x) \not\equiv 0$).

We might hope to extend the mean-value theorem to cases where the derivative does not necessarily exist, but the most obvious generalization is certainly false. For example, if $f(x) = |x|$ we have $f'_+(x) = -1$ for $x < 0$ and $f'_+(x) = 1$ for $x \geqslant 0$, so that although $f(1) = f(-1)$ we do not have $f'_+(x) = 0$ for any x at all. Still less can we expect a mean-value theorem to hold for one of the Dini derivatives. However, something analogous to the mean-value theorem still holds for Dini derivates and can substitute for the mean-value theorem is some applications.[39a]

We shall establish the following result. *Let f be continuous in $[a, b]$. If C is any number that is larger than $[f(b) - f(a)]/(b - a)$, then at uncountably many points x in (a, b) we have $f^+(x) \leqslant C$. Similarly $f^-(x) \geqslant c$ at uncountably many points x if $c < [f(b) - f(a)]/(b - a)$; not, in general, the same points in both cases.*

The proof of this proposition is much like the conventional proof of the mean-value theorem. Let k exceed $f(b) - f(a)$, and consider the function g defined by

$$g(x) = f(x) - f(a) - k\frac{x - a}{b - a}.$$

Then $g(a) = 0$ and $g(b) = f(b) - f(a) - k < 0$. Take any number s such that $0 = g(a) > s > g(b)$. Consider the set of points x in $[a, b]$ such that $g(x) \geqslant s$; this is the inverse image of a closed set, and so, since g is continuous, it is closed. Since this set is also bounded, it has a largest point, say x_s, with $g(x_s) = s$ (since g is continuous). Since $g(b) < s$, we have $x_s < b$. Since $g(x_s + h) < s$ for all sufficiently small positive h (from the way in which x_s was defined), we have $g^+(x_s) \leqslant 0$, whence

$$f^+(x_s) = g^+(x_s)\frac{k}{b-a} \leqslant \frac{k}{b-a} = C.$$

Different s's generate different x_s's, and there are uncountably many s's between 0 and $g(b)$. That is, there are uncountably many points x such that $f^+(x) \leqslant C$.

It is a familiar fact that *if f' exists and is nonnegative throughout an interval, f is nondecreasing in that interval.* This follows from the mean value theorem, since $f(y) - f(x) = (y - x)f'(c)$, with $x < c < y$. If $f'(c) \geqslant 0$ we infer that $f(y) \geqslant f(x)$ whenever $y \geqslant x$. We can use the theorem that we have just proved to establish a result that is stronger in two directions: we do not need to suppose that f' exists, and we can omit a countable number of points. More precisely, *if f is continuous and one Dini derivate is nonnegative except perhaps for a countable number of points, it follows that f is nondecreasing.*

Suppose that $f^+(x) \geqslant 0$ for $a \leqslant x \leqslant b$ except for a countable number of points. (The hypothesis $f_+(x) \geqslant 0$ implies this, and the proof for $f^-(x)$ is similar.) If f fails to be nondecreasing, there must be two points x and y such that $y > x$ and $f(y) < f(x)$. Our generalization of the mean value theorem, with $f(y) - f(x) < c < 0$, then says that there are uncountably many points in (x, y) at which $f^+ < 0$, contradicting our hypothesis.

Exercise 21.10. The continuity of f is essential for the preceding theorem: construct a discontinuous function, not nondecreasing, for which $f_+(x) \geqslant 0$ for all x.

We can now show that, *if f is continuous, all four Dini derivates have the same upper and lower bounds in any interval,* and, indeed, more generally that the collection of difference quotients $[f(x + h) - f(x)]/h$ has the same upper and lower bounds as the Dini derivates, provided, of course, that both x and $x + h$ belong to the interval in question. Suppose, for example, that $f^+(x) \geqslant m$. Let $g(x) = f(x) - mx$. Since $g^+(x) \geqslant 0$, the function g is nondecreasing. If $h > 0$, we therefore have $g(x + h) - g(x) \geqslant 0$, or in other words

$$f(x + h) - f(x) - (x + h)m + mx \geqslant 0,$$

that is,

$$\frac{f(x + h) - f(x)}{h} \geqslant m,$$

whence $f + (x) \geqslant m$. Similarly,

$$f(x) - f(x - h) - mx + (x - h)m \geqslant 0,$$

$$\frac{f(x - h) - f(x)}{-h} \geqslant m,$$

whence $f^-(x) \geqslant f_-(x) \geqslant m$.

Next, suppose that one derivate, say f^+, is continuous at x. This means that its upper and lower bounds are arbitrarily close to $f^+(x)$ in a sufficiently small neighborhood of x; the preceding theorem tells us that the same is true for the upper and lower bounds of the other three derivates. This means that all four derivates concide (with $f^+(x)$) at the point x. That is, *if one derivate of a continuous function is continuous at a point there is a derivative at that point*.

A common error of students of calculus is to suppose that $f'(y)$ cannot exist if $\lim_{x \to y} f'(x)$ does not exist. (Cf. p. 135.)

Exercise 21.11. Show that $f'(y)$ does exist if $\lim_{x \to y} f'(x)$ exists.

One reason for this misunderstanding is perhaps the fact that a derivative, if discontinuous, is very discontinuous, so that functions with discontinuous derivatives are not commonly encountered in calculus. More precisely, *a derivative cannot have a simple jump*. This is to be interpreted in the following sense: if $f'(x)$ exists at every point x of an interval, and (for a point y of this interval) the limits $f'(y+)$ and $f'(y-)$ both exist, then both these limits are equal to $f'(y)$. On the other hand, the example $f(x) = |x|$ shows that the limits of f' from both sides can exist and be different at y if $f'(y)$ does not exist. The impossibility of a simple jump for a derivative is an immediate consequence of the fact that a derivative has the intermediate value property.

A continuous function cannot have a derivative that is everywhere infinite. Indeed, we can say much more: a continuous function must have $f^+(x) < +\infty$ on an uncountable set,[40] a fact that follows at once from the generalized law of the mean on p. 141. Indeed, the generalized mean-value theorem says that $f^+(x) < C$ on an uncountable set if $C > [f(b) - f(a)]/(b - a)$.

It follows from a general theorem that we shall quote later (p. 147) that whether f is continuous or not, it can have an infinite right-hand derivative f'_+ at most on a set of measure zero. On the other hand, if we do not require f to be continuous, we can have $f^+(x) = +\infty$ at every point x. An example of this phenomenon can be constructed as follows.[41] Let real numbers x in $[0, 1]$ be represented in base 3, $x = 0.a_1 a_2 \ldots$, where each a_n is 0, 1, or 2. If x has two representations, we choose the one that terminates. Then we put $f(x) = 0.b_1 b_2 \ldots$ (base 2), where $b_n = 1$ if $a_n = 2$ and otherwise $b_n = 0$. Now, since we excluded ternary representations ending in repeated 2's, the ternary representation of every x contains an infinite sequence of digits that are 0 or 1. Let one of these 0's or 1's occur at the rth ternary place. Let x' differ from x

only by having 2 as its rth ternary digit; then $x' > x$ and in fact $x' - x = 3^{-r}$ or $2 \cdot 3^{-r}$. In either case, $f(x') - f(x) = 2^{-r}$. Hence

$$\frac{f(x') - f(x)}{x' - x} \geqslant \frac{3^r}{2^r}.$$

Since r can be arbitrarily large, it follows that $f^+(x) = +\infty$.

It can be shown that this function f is continuous except at the points that have terminating ternary expansions, and in fact is continuous on the right at these points, but discontinuous on the left.

Another interesting result about possible values of derivatives (of not necessarily continuous functions) is that if one level set of f' is dense then any other level set of f' is of first category. That is to say, if $f'(x) = A$ (possibly infinite) on a dense set E, then $f'(x)$ can exist (finite) and be different from A at most on a set of first category; hence E must be a set of first category.[41a]

It is enough to consider the set S where $f'(x) < A$, since the set where $f'(x) > A$ is the set where $(-f)'(x) < -A$. When A is finite, S is contained in the union of the sets $E_{n,m}$, where $x \in E_{n,m}$ provided that $|y - x| < 1/n$ implies

$$\frac{f(y) - f(x)}{y - x} < A - 1/m;$$

when $A = +\infty$, replace $A - 1/m$ by m. If we show that each $S_{n,m}$ is nowhere dense, we shall have proved our assertion.

Suppose then that some $E_{N,M}$ is dense in some interval I. Since the set of points where $f'(x) = A$ is dense in I, let $x_0 \in I$ be a point where $f'(x_0) = A$. Since $E_{N,M}$ is dense in I, the interval $(x_0 - 1/N, x_0 + 1/N)$ contains points of $E_{N,M}$; so we can choose $x_k \in E_{N,M}$ with $x_k \to x_0$. Thus

$$\frac{f(x_k) - f(x_0)}{x_k - x_0} < A - 1/M \quad (\text{or } < M \text{ if } A = +\infty).$$

Letting $k \to \infty$ we get $f'(x_0) \leqslant A - 1/M$ (or $\leqslant M$), contradicting $f'(x_0) = A$ (or $+\infty$). Therefore $E_{n,m}$ is always nowhere dense.

This theorem can be generalized by using Dini derivates in the hypothesis; but the proof is more complicated. Since it is easy to show that at a point of discontinuity at least one Dini derivate is infinite, it is then easy to get the following result:

If f is discontinuous at the points of an everywhere dense set and differentiable (with a finite derivative, and hence continuous) at the points of another everywhere dense set, then it must be continuous and not differentiable at the points of a set of second category.[42] We showed (p. 117) that when f is continuous at the points of an everywhere dense set, its points of discontinuity form a set of first category. Thus the presence of an everywhere dense set of points of continuity means that there are only relatively few points of discontinuity; we now see that the presence of an everywhere dense set of points of discontinuity allows only relatively few points where a derivative exists.

A direct proof can be given as follows.

Let E_n be the set of points x such that $|y - x| < 1/n$ implies $|f(y) - f(x)|/|y - x| < n$. In the present context "differentiable" means "having a finite derivative," so every point x where $f'(x)$ exists belongs to some E_n. To show that the set of such points is of first category it is then enough to show that each E_n is nowhere dense.

Suppose, on the contrary, that some E_N is dense in an open interval I. This interval contains a point w at which f is discontinuous, so there must be a positive h and a sequence $\{y_k\}$ such that $y_k \to w$ and $|f(y_k) - f(w)| \geq h$. Let k be so large that $|y_k - w| < 1/N$. Since E_N is dense in I, we can choose x_k in E_N so that x_k is between y_k and w; then $|x_k - w| < 1/N$ and $|y_k - x_k| < 1/N$. Now $h \leq |f(y_k) - f(w)| \leq |f(y_k) - f(x_k)| + |f(x_k) - f(w)|$,

$$\frac{h}{|y_k - w|} < \left| \frac{f(y_k) - f(x_k)}{y_k - w} \right| + \left| \frac{f(x_k) - f(w)}{y_k - w} \right|$$

$$\leq \left| \frac{f(y_k) - f(x_k)}{y_k - x_k} \right| + \left| \frac{f(x_k) - f(w)}{w - x_k} \right| < 2N,$$

since $x_k \in E_N$. Letting $y_k \to w$, we have a contradiction.

If one of the Dini derivates of a continuous function is zero everywhere in an interval, the function is constant there; for we have shown that the function is both nonincreasing and nondecreasing. This implies that *two continuous functions having the same finite derivative throughout an interval differ only by a constant there*. On the other hand, it is possible for two continuous functions to have the same derivative, necessarily infinite at some points, throughout an interval, and not differ by a constant there (see p. 153).

A considerable amount can be said about the derivates of a perfectly arbitrary function. We state the following facts without proof.[43] First, except at the points of a countable set, the upper derivate on one side is not less than the lower derivate on the other side. Next, if $f^+ = +\infty$ on a set E, then $f_- = -\infty$ on E except for a set of measure zero; similarly, if $f_+ = -\infty$ then $f^- = +\infty$ with the same possibility of exception. Finally, the set where f^+ and f_- are finite and different is of measure zero. Putting these facts together, we see that *except on a set of measure zero there are only three possibilities*: (1) *there is a finite derivative*; (2) *the two upper derivates are* $+\infty$ *and the two lower derivates are* $-\infty$; (3) *the upper derivate on one side is* $+\infty$, *the lower derivate on the other side is* $-\infty$, *and the other two derivates are finite and equal*. Since only (1) is possible for a monotonic function, we see in particular that a monotonic function has a finite derivative almost everywhere; we shall give a direct proof of this in the next section. Another conclusion that we can draw from the general theorem is that if all derivates are bounded almost everywhere, the function has a derivative almost everywhere.

Derivatives are not as simple functions as one might

think. For example, the product of two derivatives is not necessarily a derivative.[43a]

22. Monotonic functions. A function f from an interval I in R_1 to R_1 is called *monotonic* if it is either nondecreasing or nonincreasing. That is, f is monotonic if either $f(y) \geqslant f(x)$ whenever $y > x$ (x and y in I), or else $f(y) \leqslant f(x)$ whenever $y > x$ in I. If one of these conditions holds with strict inequality, we say that f is *strictly monotonic*. The familiar functions used in calculus are, if not monotonic, at least piecewise monotonic. Thus if $f(x) = x^2$, f is decreasing when $x < 0$ and increasing when $x > 0$; if $f(x) = \cos x$, f is alternately increasing and decreasing in the intervals $(-\pi, 0)$, $(0, \pi)$, and so on; if $f(x) = e^x$, f is increasing throughout R_1. All these functions are continuous. On the other hand, the function f defined by $f(x) = [x]$ (the greatest integer not exceeding x) is nondecreasing and has a discontinuity at each integer x.

Exercise 22.1. A monotonic function is bounded on each compact subinterval of its domain.

Exercise 22.2. A monotonic function approaches a (finite) limit from each side at every interior point of its domain.

Exercise 22.3. The limit of a pointwise convergent sequence of monotonic functions is monotonic.

A function f is said to have a *jump* at the point x of its domain if f has limits from both sides at x, but is not continuous at x. After Exercise 22.2 we can say that the only discontinuities of a monotonic function are jumps. The easiest monotonic functions to visualize are those with only a finite number of jumps, but a monotonic function can have a much more complicated structure than this. For example, if $f(x) = 2^{-n}$ in the interval $[1/(n + 1), 1/n)$, f is a nondecreasing function with jumps that have a limit point at 0.

There can be at most a countably infinite number of jumps of a monotonic function, since the intervals from $f(x -)$ to $f(x +)$, if not empty, form a set of disjoint intervals in R_1 (disjoint because f is monotonic), and such a set of intervals is countable (p. 29). However, we shall show that the set of jumps of a monotonic function can be any countable set at all, even an everywhere dense one, for example, all the rational points in an interval. Let $\{x_n\}$ be a given countable set, and let j_n be positive numbers such that $\Sigma j_n < \infty$. We define functions f_n by putting $f_n(x) = 0$ for $x < x_n$ and $f_n(x) = j_n$ for $x \geqslant x_n$. Of course, the x_n will not in general be numbered in order of increasing magnitude. The series Σf_n converges uniformly (by the *M*-test, p. 102), since $|f_n(x)| \leqslant j_n$ and Σj_n converges. If x_0 is not any of the x_n, it is a point of continuity for all the f_n and hence is a point of continuity for f (p. 103). On the other hand, if x_m is one of the x_n, precisely one function f_n, namely f_m, is discontinuous at x_m. Then $\sum_{n \neq m} f_n = f - f_m$ is continuous at x_m. Hence f, as the sum of a function that is continuous at x_m and a function that is discontinuous at x_m, is itself discontinuous at x_m. Indeed, f has a jump of amount j_m at x_m. We may reasonably call such an f a pure jump function. More generally we call f a pure jump function if it is constructed similarly, but possibly with both right-hand and left-hand jumps, so that $f(x_m -) \neq f(x_m) \neq f(x_m +)$. If we construct a pure jump function whose right-hand and left-hand jumps are just those of a given nondecreasing function g, then $g - f$ will still be nondecreasing, and also continuous.

It may seem plausible that a pure jump function should

have its derivative zero except at its jumps. This conjecture is almost, but not quite, true: *the derivative of a pure jump function is zero except on a set of measure zero*, but this set of measure zero may contain more points than the jumps.[44] We can get a better idea of the kind of thing that can happen by considering some special cases. Let f be the pure jump function that has jumps of amount 2^{-n} at the points 3^{-n}, $n = 1, 2, \ldots$; let g be the pure jump function that has jumps of amount 3^{-n} at the points 2^{-n}; let $f(0) = g(0) = 0$. Both f and g are continuous (on the right) at 0. However, we can easily show that $f'_+(0) = +\infty$ while $g'_+(0) = 0$. In fact, if $h > 0$, we have $[f(0 + h) - f(0)]/h = f(h)/h$; and if $3^{-m-1} < h < 3^{-m}$, $f(h) = \sum_{k=m+1}^{\infty} 2^{-k} = 2^{-m}$, so that $f(h)/h > 3^m/2^m \to \infty$. Similarly, if $2^{-m-1} < h < 2^{-m}$, we have $g(h) = \sum_{k=m+1}^{\infty} 3^{-k} = \frac{1}{2} \cdot 3^{-m}$, so that $g(h)/h < 2^m/3^m \to 0$.

Exercise 22.4. Construct a monotonic pure jump function, having jumps with 0 as a limit point, such that $f'_+(0)$ is positive and finite.

There seems to be no essentially simpler way of proving that a pure jump function has a zero derivative almost everywhere than to appeal to the general theorem (that we shall prove presently) that every monotonic function has a finite derivative almost everywhere.

What is perhaps more surprising is that there can be a continuous monotonic function, not a constant, whose derivative is zero almost everywhere. Functions with this property are called *singular* monotonic functions. We shall construct a singular monotonic function in some detail, since it can be used for various applications. One application (p. 153) is to the construction of a more complicated singular monotonic function which is constant in no interval.

We base the construction of a singular function on the Cantor set of §6; our example will be constant in each complementary interval of this set, and so its derivative will certainly be zero except perhaps at the points of the Cantor set, which is of measure zero. If x is any point of the interval $[0, 1]$, we write $x = 0.a_1a_2a_3 \ldots$ (base 3), so that each a_k is 0, 1, or 2. The endpoints of the complementary intervals of the Cantor set are the numbers with terminating ternary expansions that can be written without using any 1's, for example $\frac{1}{3} = 0.1000 \ldots = 0.0222 \ldots$; $\frac{2}{3} = 0.2000 \ldots$. If we halve all the digits of such an expansion and interpret the result as a number written in base 2, we get, for example, the numbers $0.0111 \ldots$ (base 2) and $0.1000 \ldots$ (base 2), which are the same. This happens for every pair of endpoints of the same complementary interval. Let us now define a function f by writing all the points x of the Cantor set in base 3, using no 1's, halving all the digits, and setting $f(x)$ equal to the resulting number, interpreted in base 2. This defines f on the Cantor set, and we have just seen that f has the same value at the two ends of each complementary interval. We can extend f to all of $[0, 1]$ by giving it the same value throughout each complementary interval that it has at the endpoints of that interval. We must now show that f is monotonic and continuous; it will also be interesting to investigate the derivates of f at the points of the Cantor set.

If x and y are two points of the Cantor set, not endpoints, and $x < y$, then the ternary expansions of x and y must be of the forms

$$x = 0.a_1a_2a_3 \ldots a_n a_{n+1} \ldots ,$$

$$y = 0.a_1a_2a_3 \ldots a_n b_{n+1} \ldots ,$$

with $b_{n+1} > a_{n+1}$. The binary expansions of $f(x)$ and $f(y)$ will then coincide through the nth digit, while the $(n + 1)$th digit of $f(y)$ will be 1 larger than the $(n + 1)$th digit of $f(x)$. This means that $f(y) > f(x)$. Hence f is a nondecreasing function.

Since f is continuous in the intervals where it is constant, we have only to consider its continuity at points of the Cantor set. Let x be such a point. A neighborhood of x contains all points y of the Cantor set which differ from x by at most 3^{-n}, that is, by numbers whose ternary expansions start with at least n zeros. The binary expansion of $f(y)$ then differs from that of $f(x)$ by a number whose binary expansion starts with at least n zeros, so that $f(y)$ differs from $f(x)$ by at most 2^{-n}. Since for a point y not in the Cantor set but in the same neighborhood of x the value of $f(y)$ is the same as its value at either endpoint of the complementary interval containing y, it follows that f is continuous at x.

We have therefore shown that f is continuous, monotonic, not constant, and singular. We now investigate the differentiability of f at points of the Cantor set. At a left-hand endpoint of a complementary interval, the right-hand derivative f'_+ exists and is 0; and similarly $f'_-(x) = 0$ at a right-hand endpoint x.

Consider first the derivatives on the other sides of endpoints of complementary intervals, say for definiteness a right-hand endpoint $x = 0.a_1 a_2 \ldots a_n 20000 \ldots$. If h is between 3^{-m} and 3^{-m-1}, and $m > n$, $f(x + h)$ differs from $f(x)$ by something between 2^{-m} and 2^{-m-1}, so that $[f(x + h) - f(x)]/h$ must be between $2^{-m-1}/3^{-m}$ and $2^{-m}/3^{-m-1}$. Hence as $h \to 0$ (and so $m \to \infty$), this difference quotient becomes positively infinite. That is, at a right-hand endpoint x, we have $f'_+(x) = +\infty$ and $f'_-(x) = 0$. Similarly, $f'_-(x) = +\infty$ and $f'_+(x) = 0$ at a left-hand endpoint.

At limit points that are not endpoints it can be shown that $f^+ = +\infty$ whereas f_+ can have any value between 0 and $+\infty$.

As a first application we construct another singular function, which is not constant on any interval.[45] Let C be the function just constructed, but extended to $(-\infty, \infty)$ by putting $C(x) = 0$ for $x \leqslant 0$ and $C(x) = 1$ for $x \geqslant 1$.

Let $\{r_n\}$ be an enumeration of the rational numbers. Define

$$f(x) = \sum_{n=1}^{\infty} 2^{-n} C(2^n(x - r_n)).$$

Since the series is uniformly convergent (M-test), f is continuous. Also f is strictly increasing since if $x < y$ there is a rational r between x and y. Then $C(2^n(x - r)) = 0 < C(2^n(y - r))$. For each $m \neq n$,

$$C(2^m(x - r)) \leqslant C(2^m(y - r))$$

because C is nondecreasing. Hence $f(x) < f(y)$.

Finally, by Fubini's theorem on differentiation of series with nondecreasing terms (see p. 160, below),

$$f'(x) = \sum_{n=1}^{\infty} C'(2^n(x - r_n)) = 0$$

for almost all x.

As another application of the singular function C constructed on p. 151, we can construct the example, mentioned on p. 147, of two functions that have the same derivatives (infinite at some points) throughout an interval, but do not differ by a constant. We also need a function g that is continuous and nondecreasing with a finite derivative at each point not in the Cantor set and derivative $+\infty$ at each point of the Cantor set. Once we have such a g, we can put $h(x) = f(x) + g(x)$; then

$h'(x) = g'(x) = +\infty$ at all points of the Cantor set (since all derivates of f are nonnegative); and $h'(x) = g'(x)$ at all points not in the Cantor set, since $f'(x) = 0$ at such points. However, g and h differ by f, which is not constant.

We proceed to construct g.[46] Let us enumerate the complementary intervals (a_n, b_n) of the Cantor set in order of decreasing length (the order among the finitely many intervals of each length is irrelevant). Let φ_n be a

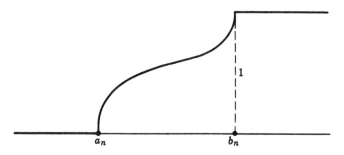

continuous nondecreasing function of the general character indicated in the figure above: $\varphi_n(x) = 0$ for $x < a_n$, $\varphi_n(x) = 1$ for $x > b_n$, $\varphi_{n+}'(a_n) = \varphi_{n-}'(b_n) = +\infty$. (An explicit example is

$$\varphi_n(x) = (2/\pi)\tan^{-1}\{(x - a_n)^{1/2}(b_n - x)^{-1/2}\}.)$$

We now observe that the lengths of the intervals (a_n, b_n) are negative integral powers of 3; let $h_n = (\frac{2}{5})^m$ whenever (a_n, b_n) has length 3^{-m}. Define $g(x) = \Sigma h_n \varphi_n(x)$ with the (infinite) sum extended over all n. In other words, the value $g(x)$ is the sum of the h_n over all intervals (a_n, b_n) to the left of x, plus $h_k \varphi_k(x)$ if x is in (a_k, b_k). Now there are 2^{m-1} intervals (a_n, b_n) of length 3^{-m}, and $h_n = (\frac{2}{5})^m$ on each, so Σh_n converges. Therefore the series defining g converges uniformly and so g is continuous. By construction, g is nondecreasing. Let x be a point of the Cantor set, other than an a_n, and let $\delta > 0$. The difference quotient

$$\Delta = \delta^{-1}\{g(x + \delta) - g(x)\}$$

$$= \delta^{-1}\Sigma h_n\{\varphi_n(x + \delta) - \varphi_n(x)\}$$

will exceed h_k if the interval $(x, x + \delta)$ contains the complementary interval (a_k, b_k). It is easy to see that $(x, x + \delta)$ always contains a complementary interval of length $3^{-m} \geqslant \delta/9$. Then $\Delta \geqslant (\frac{2}{5})^m/\delta \geqslant \frac{1}{9}3^m(\frac{2}{5})^m \to \infty$, so $g'_+(x) = +\infty$. If, however, x is an a_n then $g'_+(x) = +\infty$ by inspection. Similarly $g'_-(x) = +\infty$ at all x in the Cantor set. That is, $g'(x) = +\infty$ at all x in the Cantor set.

We now turn to the rather difficult proof that *a monotonic function has a finite derivative almost everywhere.*[47] The reader who is interested in seeing some applications of the theorem first may skip to p. 160.

The proof depends on a lemma by F. Riesz, known as the "flowing water" or "rising sun" lemma. If g is a continuous function from an interval I to R_1, if the graph of g is the cross section of the bed of a stream, and we consider the set E of points where water is flowing, it is intuitively clear that E consists of open intervals at whose

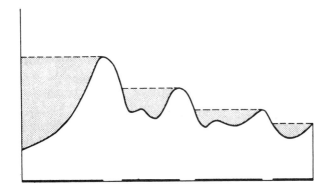

ends g has the same value; if the graph is the profile of a mountain range, if the sun rises in the direction of the positive x-axis, and if E is the set of points that are in the shade, it is again intuitively clear that E consists of open intervals at whose ends g has the same value. (In both cases there may be an exceptional interval at the extreme left, as in the picture.)

We now state the lemma in abstract terms and for a more general situation. *Let g be continuous on an interval I, except for jumps, and let G be defined by*

$$G(x) = \max(\, g(x-\,), g(x), g(x+\,)).$$

The set E of points x such that there is a $y > x$ with $g(y) > G(x)$ is an open set; and if (a,b) is any one of the intervals making up E, then $g(a+\,) \leqslant G(b)$.

If we exchange left and right, and define E' as the set of points x such that there is a $y < x$ with $g(y) > G(x)$, then, similarly, if E' is the union of open intervals (a',b') we have $G(a') \geqslant g(b'-\,)$.

We first prove that E is open. Let $x_0 \in E$; then there is a $y > x_0$ with $g(y) > G(x_0)$. We have to show that this property holds for all x near x_0. If x varies slightly to the left of x_0, $g(x_0)$ is near $g(x-\,)$; if x varies slightly to the right of x_0, $g(x_0)$ is near $g(x+\,)$; in either case $G(x)$ can exceed $G(x_0)$ only slightly. Since $g(y) > G(x_0)$, we also have $g(y) > G(x)$ as long as $G(x)$ is not much greater than $G(x_0)$, which, as we have just seen, is the case when x is near x_0.

To prove the second statement of the lemma, let (a,b) be one of the intervals that make up E, so that b is not a point of E. Let $a < x < b$; it will be enough to show that $g(x) \leqslant G(b)$, and then let $x \to a+$. Since $x \in E$ there are points $y > x$ with $g(y) > G(x)$, and hence with $G(y) \geqslant g(x)$. We wish to show that b is such a point y. If it is not, we have $G(b) < g(x)$. Let z_1 be the least upper bound of points z in $[x, b]$ such that $G(z) \geqslant g(x)$. (There are such points, for instance, x; we are trying to show that $z_1 = b$.) If z_1 is not b, then z_1 belongs to E and there is accordingly a $y > z_1$ with $g(y) > G(z_1)$. If this y is in $(z_1, b]$ we have $G(y) < g(x) \leqslant G(z_1)$, contrary to the way in which z_1 was defined; so $y > b$. But then since b is not in E, $g(y) \leqslant G(b)$. Hence if $G(b) < g(x)$, we have

$$g(x) \leqslant G(z_1) < g(y) \leqslant G(b) < g(x),$$

a contradiction.

We are going to deduce the differentiability of a nondecreasing function from two consequences of Riesz's lemma. We lead up to these by showing first that our theorem will follow if we can prove that $f^+(x) < +\infty$ almost everywhere and $f^+(x) \leqslant f_-(x)$ almost everywhere. It will simplify the notation somewhat, and will do no harm, if we suppose that our interval (a, b) has its midpoint at 0. We can then reflect the graph of f around the origin to obtain the graph of the function f_0 whose value at x is $-f(-x)$. This function f_0 is also nondecreasing and has its lower derivate on the right at x equal to the lower derivate of f on the left at $-x$; in fact,

$$\frac{f_0(x + h) - f_0(x)}{h} = \frac{-f(-x - h) - (-f(-x))}{h}$$

$$= \frac{f(-x + (-h)) - f(-x)}{-h}.$$

If $f^+(x) \leqslant f_-(x)$ holds almost everywhere for every nondecreasing f, it holds also with f replaced by f_0 and x replaced by $-x$. That is, $f^-(x) \leqslant f_+(x)$ for almost all x, so that for almost all x we have

$$f^+(x) \leqslant f_-(x) \leqslant f^-(x) \leqslant f_+(x) \leqslant f^+(x).$$

If $f^+(x) < \infty$ for almost all x the preceding inequality shows that all four derivates are equal and finite for almost all x, that is, that there is a finite derivative for almost all x.

Next we can reduce the problem still further; it is enough to show that the set on which $f_-(x) < r < R < f^+(x)$ has measure zero, whatever r and R may be. For, if $f^+(x) > f_-(x)$, there are rational numbers r and R such that $f_-(x) < r < R < f^+(x)$. Since there are only countably many pairs of rational numbers, the set where $f^+(x) > f_-(x)$ is contained in the union of countably many sets of measure zero and so is itself of measure zero.

The consequences of Riesz's lemma on which our proof depends are as follows:

(1) Let f be nondecreasing on $[a, b]$ and let E_R be the set of points x where f is continuous and $f^+(x) > R$. Then E_R can be covered by a countable set of intervals (a_k, b_k) whose total length $\Sigma(b_k - a_k)$ is at most

$$R^{-1}\Sigma\big[f(b_k +) - f(a_k +)\big] \leqslant \big[f(b -) - f(a +)\big]/R.$$

(2) Let f be nondecreasing on $[a, b]$ and let E_r be the set of points x where f is continuous and $f_-(x) < r$. Then E_r can be covered by a countable set of intervals (a_k, b_k) such that $\Sigma[f(b_k -) - f(a_k +)] \leqslant r\Sigma(b_k - a_k) \leqslant r(b - a)$.

We defer the proofs of these two statements until we have shown how the theorem follows from them.

As a first step, we observe that (1) implies $f^+(x) < +\infty$ almost everywhere. For, if $f^+(x) = +\infty$ on E we have the hypothesis of (1) for every positive R, so that E is covered by intervals (a_k, b_k) whose total length is at most $[f(b -) - f(a +)]/R$ for every R. That is, E is covered by intervals of arbitrarily small total length, whence E is of measure zero.

Next consider a set E where f is continuous and $f_-(x) < r < R < f^+(x)$, so that the hypotheses of (1) and (2) are both satisfied. Apply (1) to each of the intervals (a_k, b_k) in (2). We find that the part of E in (a_k, b_k) is covered by intervals whose total length, L_k say, is at most $[f(b_k -) - f(a_k +)]/R$. Adding these inequalities for all (a_k, b_k), and applying (2), we find that

$$\Sigma L_k \leqslant (1/R)\Sigma[f(b_k -) - f(a_k +)] \leqslant (r/R)(b - a).$$

The same argument applies to any subinterval of (a, b); that is, the part of E in any interval (p, q) is covered by intervals of total length at most $(r/R)(q - p)$. Now let E be covered in any way by subintervals (p_k, q_k) (overlapping or not). The part of E in (p_k, q_k) can be covered by nonoverlapping intervals of total length at most (r/R) $(q_k - p_k)$, and so E can be covered by intervals whose total length is at most $(r/R)\Sigma(q_k - p_k)$. The existence of such coverings implies that E is of measure zero (Exercise 11.1).

We now turn to the proof of propositions (1) and (2).

(1) If $x \in E_R$ there is a $y > x$ such that $[f(y) - f(x)] \div (y - x) > R$, that, $f(y) - Ry > f(x) - Rx$. Apply Riesz's lemma to the function g such that $g(x) = f(x) - Rx$. Since f is continuous at x, so is g, and $G(x) = g(x)$. We conclude that E_R is covered by a countable set of intervals (a_k, b_k) such that $G(b_k) \geqslant g(a_k +)$, in other words (since f is nondecreasing),

$$f(b_k +) - Rb_k \geqslant f(a_k +) - Ra_k,$$

that is,

$$f(b_k +) - f(a_k +) \geqslant R(b_k - a_k).$$

Adding these inequalities for the various values of k, and again using the fact that f is nondecreasing, we have

$$R\Sigma(b_k - a_k) \leqslant \Sigma\big[f(b_k +) - f(a_k +)\big].$$

(2) If $x \in E_r$, there is a $y < x$ such that $[f(y) - f(x)] \div (y - x) < r$, that is (since $y < x$),

$$f(y) - f(x) > (y - x)r,$$
$$f(y) - ry > f(x) - rx.$$

Riesz's lemma, in the form with the inequality reversed in its hypothesis, can be applied to the function g defined by $g(x) = f(x) - rx$. We find that E_r is covered by a set of intervals (a_k, b_k) such that $g(b_k -) \leqslant G(a_k)$; since f is nondecreasing, the existence of this covering implies

$$f(b_k -) - rb_k \leqslant f(a_k +) - ra_k,$$

that is,

$$f(b_k -) - f(a_k +) \leqslant r(b_k - a_k).$$

Summing over all the intervals, we obtain the conclusion of (2).

We now give some interesting applications of the theorem on differentiating monotonic functions. The existence of such applications helps to justify the amount of effort expended on the proof of the theorem.

(i) DIFFERENTIATION OF SERIES OF MONOTONIC FUNCTIONS (FUBINI'S THEOREM).[48] *Let $f_1 + f_2 + \cdots$ be a pointwise convergent series of nondecreasing functions on an interval $[a, b]$, with sum s. Then for almost every x on $[a, b]$ we have $f_1'(x) + f_2'(x) + \cdots = s'(x)$.*

This is one example of a theorem that is more conveniently stated in terms of series than in terms of sequences.

If we exclude a set of measure zero, all the f_n have nonnegative derivatives, and so does s, since it too is a nondecreasing function. The series $f_1'(x) + f_2'(x) + \cdots$ has nonnegative terms and so its partial sums $s_n'(x)$ form a nondecreasing sequence (for each x). It therefore converges if its partial sums are bounded. But

$$\frac{s(x+h) - s(x)}{h} = \frac{f_1(x+h) - f_1(x)}{h} + \cdots$$

$$\geqslant \frac{f_1(x+h) - f_1(x)}{h} + \cdots$$

$$+ \frac{f_n(x+h) - f_n(x)}{h},$$

since the f_n are nondecreasing. Letting $h \to 0$, we infer that $s'(x) \geqslant s_n'(x)$ whenever the left-hand side is finite, that is, almost everywhere. Hence the differentiated series converges for each x and we have only to verify that it has the right sum.

To identify the sum of the differentiated series, we first show that some subsequence of its partial sums converges to $s'(x)$ almost everywhere. We observe that $s(b) - s_n(b) \to 0$, so that there must be a subsequence of integers n for which $\Sigma[s(b) - s_n(b)]$ converges (we can pick an n that makes the difference less than $\frac{1}{2}$, then a larger n that makes it less than $\frac{1}{4}$, and so on). For values of n in this subsequence we have $s(x) - s_n(x) \leqslant s(b) - s_n(b)$ since $s(x) - s_n(x)$ is the "tail" of the series for $s(x)$ and so defines a nondecreasing function. Hence $\Sigma[s(x) - s_n(x)]$ (summed over the same values of n as before) is a convergent series of nondecreasing functions. By what we have already proved, the series $\Sigma[s'(x) - s_n'(x)]$ converges almost everywhere, and consequently its general term approaches zero almost everywhere. That is, we have found a subsequence of partial sums $s_n(x)$ such that $s_n'(x) \to s'(x)$ almost everywhere. Since the whole sequence of partial sums converges almost everywhere, it must converge to the limit of the subsequence, that is, to $s'(x)$.

(ii) DENSITY OF SETS. Let E be a set in R_1. We say that a point x (in E or not) is a *point of density* for E if sufficiently small neighborhoods of x consist "mostly" of points of E. It is not quite easy to formulate this definition precisely. First let us cover E by a countable collection of open intervals. This can be done in various ways; we take the greatest lower bound of the total lengths of such coverings as giving a measure of the size of E, and call it the *exterior measure* of E, denoted by $\mu(E)$. For an interval I, $\mu(I)$ is the ordinary length of I. If E_1 and E_2 are sets lying in two disjoint intervals, $\mu(E_1 \cup E_2) = \mu(E_1) + \mu(E_2)$. Now let N stand for a neighborhood of x; we say that x is a point of density for E if $\mu(E \cap N)/\mu(N) \to 1$ as $\mu(N) \to 0$; in other words, if the exterior measure of the part of E in a small neighborhood of x is nearly equal to the length of the neighborhood. There is a similar definition for sets in any R_n.

We shall now prove that *almost all points of E are points of density for E*. This means that, generally speaking, a set nearly fills small neighborhoods of its points, and cannot, for example, occupy about half of every interval. (Compare Exercise 11.1. The same theorem also holds in any R_n.)

We suppose that E is not of measure zero; otherwise the theorem has no content. We may also suppose that E is bounded, and hence lies in some compact interval I, since only small neighborhoods of points of E are relevant. Define a function λ by putting $\lambda(x)$ equal to the exterior measure of the part of E that is to the left of x. Then λ is a nondecreasing function and our theorem amounts to showing that $\lambda'(x) = 1$ for almost all x in E.

Let us first consider a function f analogous to λ but obtained by using some fixed covering of E by a countable set of intervals; that is, $f(x)$ is the total length of intervals belonging to the covering and lying to the left of x; if x is inside one or more of the intervals, we count only the part of each interval that is to the left of x. Now $f'(x) = 1$ when $x \in E$, since if $x \in E$ and h is small enough, $x + h$ will be in one of the (open) intervals covering x, and so $f(x + h) - f(x) = h$. Take a sequence of coverings of E whose total lengths μ_n approach $\mu(E)$ rapidly enough so that $\Sigma[\mu_n - \mu(E)]$ converges. Let f_n be the function f associated with the nth covering. Then $f_n(x) \to \lambda(x)$, and $f_n(x) - \lambda(x) \leqslant \mu_n - \mu(E)$; moreover, $f_n - \lambda$ is a nondecreasing function. We have $\Sigma[f_n(x) - \lambda(x)]$ convergent, and by Fubini's theorem the differentiated series $\Sigma[f_n'(x) - \lambda'(x)]$ converges almost everywhere. Hence the terms of this series approach zero almost everywhere. Since $f_n'(x) = 1$ almost everywhere in E, we must also have $\lambda'(x) = 1$ almost everywhere in E.

(iii) THE MEASURE OF A LOCUS.[49] Let F be a compact set in R_1. If x is not in F we know that there is a positive distance from x to F and that it is attained for some point in F (Exercise 8.9). *Consider the set E_r of points that are at distance r from F, where $r > 0$. Then E_r is a set of measure zero.* For, if not, E_r contains a point of density; let y be such a point. Let x be a point in F at distance r from y. A neighborhood N of x of radius r cannot contain any point of E_r, since all points of N are distant less than r from y. Now any neighborhood I of y is half in N and so

$I \cap C(E_r)$ contains an interval at least half as long as I. The existence of such intervals contradicts the assumption that y is a point of density for E_r.

23. Convex functions. We consider functions from an interval in R_1 to R_1. A function is usually called *convex* if the portion of its graph in every interval lies on or below its chord. We shall establish the rather striking fact that a continuous function is convex if we merely assume that the midpoint of each chord is above (or on) the graph of the function. Actually much milder hypotheses than continuity suffice; for example, it is enough to suppose that the function is bounded in some (small) interval.[50] Hence if a discontinuous function has the midpoint property, its graph must be rather wild. There exist such functions, for example, the discontinuous linear functions mentioned in §20.

The geometrical statements about chords can be formulated analytically as follows: to say that the midpoint of every chord lies on or above the graph means that

$$f\left(\frac{x+y}{2}\right) \leqslant \frac{f(x) + f(y)}{2}$$

for every x and y in the domain; to say that the entire chord lies on or above the graph means that

$$f(q_1 x + q_2 y) \leqslant q_1 f(x) + q_2 f(y)$$

whenever $q_1 \geqslant 0$, $q_2 \geqslant 0$, and $q_1 + q_2 = 1$. We are to show that when f is continuous, the first inequality (for all x and y) implies the second.

It is convenient to write the first inequality in a somewhat different form. Put $\Delta_h f(x) = f(x + h) - f(x)$, so that

$$\Delta_h \Delta_k f(x) = f(x + h + k) - f(x + h) - f(x + k) + f(x).$$

Then we write

$$\Delta_h \Delta_h f(x) = \Delta_h^2 f(x) = f(x + 2h) - 2f(x + h) + f(x),$$

and the midpoint inequality says that $\Delta_h^2 f(x) \geqslant 0$ whenever x and $x + 2h$ are both in the domain of f. (Here h may be either positive or negative.)

The second inequality can be put in the form

$$\frac{\Delta_g f(x)}{g} \leqslant \frac{\Delta_h f(x)}{h}, \qquad 0 < g < h, \qquad (*)$$

as we can see by taking (with $y > x$) $z = q_1 x + q_2 y$, $h = y - x$, $g = z - x$. This says that if we fix the left-hand end of a chord, its slope is an increasing function of its length. To show that a continuous function that satisfies the midpoint definition of convexity satisfies the other definition, we have to deduce $(*)$ from $\Delta_h^2 f(x) \geqslant 0$. To do this,[51] we start from the identity

$$\Delta_h f(x) = \sum_{i=0}^{n-1} \Delta_{h/n} f(x + ih/n),$$

where n is any positive integer (the sum "telescopes"). If we apply the operator $\Delta_{h/n}$ to both sides, we obtain (if $x + h + h/n$ is still in the domain of the function)

$$\Delta_{h/n} \Delta_h f(x) = \sum_{i=1}^{n} \Delta_{h/n}^2 f(x + ih/n) \geqslant 0.$$

That is,

$$\Delta_h f(x + h/n) - \Delta_h f(x) \geqslant 0,$$

or, in other words,

$$\Delta_h f(x + h/n) \geqslant \Delta_h f(x).$$

Since this inequality holds for all x such that all points entering the formulas are in our interval, we may replace x successively by $x + h/n, x + 2h/n, \ldots,$ and so we find that for any positive integer m,

$$\Delta_h f(x + mh/n) \geqslant \Delta_h f(x + (m - 1)h/n)$$
$$\geqslant \cdots \geqslant \Delta_h f(x + h/n) \geqslant \Delta_h f(x).$$

If now x, $x + \delta$, and h are given numbers, we can find sequences of rational numbers m/n such that $mh/n \to \delta$. We then use the continuity of f to deduce

$$\Delta_h f(x + \delta) \geqslant \Delta_h f(x)$$

from

$$\Delta_h f(x + mh/n) \geqslant \Delta_h f(x).$$

We have now shown that $\Delta_h f(x)$ increases with x for each h. In particular,

$$\Delta_{h/n} f(x) \leqslant \Delta_{h/n} f(x + h/n) \leqslant \cdots$$
$$\leqslant \Delta_{h/n} f(x + (n - 1)h/n).$$

Let $0 < m < n$. The average of the first m terms of this chain of inequalities does not exceed the average of the first n terms. That is,

$$\frac{\Delta_{h/n} f(x) + \Delta_{h/n} f(x + h/n) + \cdots + \Delta_{h/n} f(x + (m - 1)h/n)}{m}$$
$$\leqslant \frac{\Delta_{h/n} f(x) + \cdots + \Delta_{h/n} f(x + (n - 1)h/n)}{n}.$$

Since the numerators again telescope, this is the same as

$$\frac{f(x + mh/n) - f(x)}{m} \leqslant \frac{f(x + h) - f(x)}{n},$$

or, if $h > 0$,

$$\frac{f(x + mh/n) - f(x)}{mh/n} \leqslant \frac{f(x + h) - f(x)}{h}.$$

If $0 < g < h$, we can choose a sequence of rational fractions m/n such that $m/n \to g/h$, and we infer that (∗) holds.

From (∗), without any initial hypothesis about the continuity of f, we can deduce not only that *f is continuous at each point in the interior of its interval of convexity*, but that *it has finite right-hand and left-hand derivatives* at such points, these derivatives themselves being nondecreasing functions. Since a monotonic function is continuous except for countably many jumps, the right-hand derivative of a convex function in sense (∗) is continuous except perhaps on a countable set. This means, in particular, that the convex function is itself continuous. In fact, the only points at which this does not follow from the existence of a derivative are the points where the right-hand and left-hand derivatives are different. At such a point the function is continuous on each side, so it has at worst a simple jump; but at a simple jump a function cannot have finite derivatives on both sides.

To deduce our statements about derivatives from (∗) we observe that (∗) says that $\Delta_h f(x)/h$ defines, for each x, an increasing function of h. As $h \to 0 +$ this ratio therefore approaches a limit, which as far as we know up to now may be either finite or infinite. That is, $f'_+ (x)$ exists (finite or infinite) for every x that is interior to our interval.

By working with negative h we infer similarly that $f'_-(x)$ exists (finite or infinite).

Observe that we have now shown that, for positive h, the quantity $h^{-1}\Delta_h f(x)$ increases with x for fixed h, and increases with h for fixed x (as is geometrically obvious). Suppose that $0 < g < h$ and $x < y$. Then by applying these two facts we obtain

$$\frac{\Delta_g f(x)}{g} \leqslant \frac{\Delta_h f(x)}{h} \leqslant \frac{\Delta_h f(y)}{h}. \qquad (\substack{* \\ *})$$

Letting $g \to 0$ with h fixed, we see that $f'_+(x) < +\infty$. Similarly $f'_-(x) > -\infty$. Moreover,

$$\frac{\Delta_{-g} f(x)}{-g} = \frac{f(x-g) - f(x)}{-g} = \frac{f(x) - f(x-g)}{g}$$

$$= \frac{\Delta_g f(x-g)}{g} \leqslant \frac{\Delta_g f(x)}{g},$$

and so $f'_-(x) \leqslant f'_+(x)$. Finally, since

$$\frac{\Delta_h f(x)}{h} \leqslant \frac{\Delta_h f(y)}{h}, \qquad y > x,$$

we infer that f'_+ is nondecreasing, and similarly f'_- is nondecreasing.

Returning to $(\substack{* \\ *})$, we also see that

$$f'_+(x) \leqslant \frac{\Delta_h f(x)}{h} = \frac{f(x+h) - f(x)}{h}$$

for every positive h. The right-hand side is the slope of an arbitrary chord of the graph of $y = f(x)$ with left-hand end at $(x, f(x))$; and the left-hand side is the slope of the right-hand tangent to the graph at x. Thus every chord through $(x, f(x))$ and going to the right lies above (or on) the (right-hand) tangent, which means that the entire curve, from x onward to the right, lies above (or on) the right-hand tangent. Similarly on the left; and since the right-hand tangent has larger slope than the left-hand tangent, the entire curve lies above (or on) the right-hand tangent line at x.

A straight line such that the entire curve lies above or on it is called a supporting line; the existence of a supporting line at each point can be (and often is) taken as the definition of convexity. We say that the function is strictly convex if each supporting line has just one point of contact with the graph.

Let us note that the usual calculus criterion for convexity is indeed a sufficient condition. Suppose that $f''(x)$ exists and is nonnegative at every point of an interval. Then for x in the interior of the interval, and small positive h, we have, by two applications of the law of the mean,

$$f(x + 2h) - f(x + h) = hf'(x + c_1),$$
$$x + h < c_1 < x + 2h;$$
$$f(x + h) - f(x) = hf'(x + c_2), \qquad x < c_2 < x + h;$$
$$\Delta^2 f(x) = h[f'(x + c_1) - f'(x + c_2)]$$
$$= h(c_1 - c_2)f''(c_3) \geqslant 0.$$

A similar argument applies when $h < 0$. Thus f is convex according to the midpoint definition.

We can now generalize the original definition of convex functions: not only is every chord of the graph above or on its arc, but if we put arbitrary positive weights at n points of the arc, their center of gravity will also be above or on the arc; and for a strictly convex function (and

$n > 1$) the center of gravity will be strictly above the arc. Algebraically, this means that if $w_1, w_2, \ldots, w_n$ are positive weights whose sum is 1, then

$$f(w_1 x_1 + w_2 x_2 + \cdots + w_n x_n)$$
$$\leqslant w_1 f(x_1) + \cdots + w_n f(x_n), \qquad (\dagger)$$

with strict inequality if f is strictly convex and at least two x_k are different. (In comparing (†) with our earlier statements about convexity we have to write x_1, x_2 for what were formerly x, y.) If f is concave, the inequality is reversed.

Inequality (†) is known as Jensen's inequality. To prove it, let $M = w_1 x_1 + w_2 x_2 + \cdots + w_n x_n$, $w_1 + \cdots + w_n = 1$, so that M is the x-coordinate of the center of gravity of n weights w_k placed at the points $(x_k, f(x_k))$. Let us consider the supporting line determined by the right-hand tangent to the graph of f at $x = M$; the curve lies above this tangent. Let the equation of the tangent be $y = g(x) = ax + b$, where a and b are numbers which could be calculated but whose values are irrelevant.

Since the curve is above or on this supporting line through $(M, f(M))$, we have $ax + b \leqslant f(x)$. Write this inequality for $x = x_1, x_2, \ldots, x_n$, multiply the kth inequality by w_k, and add the results. We get

$$w_1 g(x_1) + \cdots + w_n g(x_n) \leqslant w_1 f(x_1) + \cdots + w_n f(x_n),$$

or, since the sum of the w's is 1,

$$a + b \sum_{k=1}^{n} w_k x_k \leqslant \sum_{k=1}^{n} w_k f(x_k).$$

But $\sum_{k=1}^{n} w_k x_k = M$, so

$$g(M) \leqslant \sum_{k=1}^{n} w_k f(x_k).$$

Finally, the supporting line $y = g(x)$ goes through the point $(M, f(M))$, so $g(M) = f(M)$ and

$$f(M) \leqslant \sum_{k=1}^{n} w_k f(x_k);$$

this is (†).

Furthermore, if any x_k is different from M there was strict inequality in one of the inequalities that we added. Hence there is strict inequality in (†) unless all the x_k are equal to M.

In a similar way we can get analogous inequalities for integrals instead of sums. Let w and x be continuous positive functions, $a \leqslant t \leqslant b$, with $\int_a^b w(t)\, dt = 1$. If we replace the number M by

$$M = \int_a^b x(t) w(t)\, dt$$

and use the inequality $g(y) \leqslant f(y)$ for every y between a and b, we get

$$\int_a^b w(t) f(x(t))\, dt \geqslant f\left(\int_a^b w(t) x(t)\, dt \right), \quad \int_a^b w(t)\, dt = 1,$$

provided that f is convex, and the reversed inequality when f is concave. This is the integral form of Jensen's inequality.

The ordinary average, or more formally the arithmetic mean, of n numbers $x_1, \ldots, x_n$ is

$$\frac{x_1 + \cdots + x_n}{n} \, ;$$

the geometric mean is

$$(x_1 x_2 \cdots x_n)^{1/n}.$$

For many purposes it is better to use weighted means,

$$w_1x_1 + w_2x_2 + \cdots + w_nx_n \quad \text{and} \quad x_1^{w_1}x_2^{w_2} \cdots x_n^{w_n},$$

where the w_k add to 1; the ordinary case has $w_k = 1/n$ for each k.

A famous (and very useful) theorem states that the geometric mean of n positive numbers does not exceed their arithmetic mean and is actually less than it unless all the x_k are equal to each other. This is just a consequence of the convexity of $-\log t$ for positive t. In fact, by (†) applied to this function,

$$-\log(w_1x_1 + \cdots + w_nx_n) \leqslant \sum_{k=1}^{n} w_k \log x_k,$$

that is,

$$\log(w_1x_1 + \cdots + w_nx_n) \geqslant \sum_{k=1}^{n} \log(x_k)^{w_k}.$$

If we exponentiate both sides we get the inequality between the means; and there is equality if and only if all the x_k are equal.

There are many applications of Jensen's inequality in general and of the inequality between the geometric and arithmetic means in particular. We can, for example, solve many maximum and minimum problems for polynomials without using calculus, for example, to find the largest box that can be made from a square of paper a units on a side by cutting x by x squares out of the corners and folding up the resulting rectangles.[51a] This problem asks us to maximize $x(a - 2x)^2$, but it is just as satisfactory to maximize $\{x(a - 2x)^2\}^{1/3}$. Consider the numbers x and $a - 2x$ with weights $\frac{1}{3}, \frac{2}{3}$. Their weighted geometric mean is $x^{1/3}(a - 2x)^{2/3}$. This does not exceed the weighted arithmetic mean, namely $\frac{1}{3}x + \frac{2}{3}(a - 2x)$, and there is equality if and only if $x = a - 2x$, that is, $x = a/3$. This says that the largest possible value for $\{x(a - 2x)^2\}^{1/3}$ is attained at $x = a/3$; and for this x we have $x(a - 2x) = a^3/27$. Notice that it was important to know when equality is attained in the inequality between the means.

As another application we deduce a necessary condition for the convergence of a series $\sum a_n$ of positive terms.[52] Of course $a_n \to 0$ is a necessary condition, but $na_n \to 0$ is (in general) not. However, if the series is convergent, then n times the geometric mean of the first n terms must approach zero; that is $n(a_1 a_2 \cdots a_n)^{1/n} \to 0$.

To see this, let $s_n = a_1 + a_2 + \cdots + a_n$ and $G_n = (a_1 a_2 \cdots a_n)^{1/n}$. Then we are assuming $s_n \to s$. It follows that the arithmetic mean of $s_1, s_2, \ldots, s_n$ approaches s.

Exercise 23.1. Prove the preceding statement.

Written in terms of the a_k, this says that

$$n^{-1}\{a_1 + (a_1 + a_2) + \cdots + (a_1 + a_2 + \cdots + a_n)\} \to s,$$

that is,

$$n^{-1}\{na_1 + (n - 1)a_2 + \cdots + a_n\} \to s,$$

or equivalently

$$n^{-1}\{(n + 1)s_n - a_1 - a_2 - 2a_2 - \cdots - na_n\} \to s.$$

Since $s_n \to s$, the preceding formula shows that

$$\frac{a_1 + 2a_2 + \cdots + na_n}{n} \to 0.$$

The left-hand side is the arithmetic mean of the n numbers displayed in the numerator, so their geometric mean also approaches 0. But this geometric mean is

$$\{n!a_1 a_2 \cdots a_n\}^{1/n} = (n!)^{1/n} G_n.$$

Since $(n!)^{1/n}/n \to 1/e$ (for example, by Stirling's formula), $nG_n \to 0$.

An application of the integral form of Jensen's inequality is a property of convex functions known as Hadamard's theorem: if f is continuous and convex on (a,b) then

$$f\left(\frac{a+b}{2}\right) \leqslant \frac{1}{b-a} \int_a^b f(t)\, dt.$$

We have only to apply Jensen's inequality with $w(t) = 1/(b-a)$ and $x(t) = t$:

$$\frac{1}{b-a} \int_a^b f(t)\, dt \geqslant f\left(\int_a^b \frac{1}{b-a}\, t\, dt\right) = f\left(\frac{1}{b-a}\, \frac{b^2-a^2}{2}\right)$$

$$= f\left(\frac{b-a}{2}\right).$$

We can interpret Hadamard's inequality in physical terms by taking $f(t)$ to be the velocity of a moving point at time t. Suppose a point moves on a line with ever-increasing acceleration from time 0 to time T. Then its velocity at mid-time ($t = T/2$) cannot exceed its average velocity for the whole trip. (This may seem physically plausible, but it is true, for all continuous convex f, only for $T/2$; that is, not necessarily for $t = aT$ with $a \neq \frac{1}{2}$.)

For a less frivolous application of Jensen's inequality, take[52a] $f(x) = x^r$; when $r > 1$, f is convex, and when $0 < r < 1$, f is concave; f is also convex when $r < 0$, but the resulting inequalities do not seem to be of much use.

For $r > 1$ we have (with all sums from 1 to n)

$$\left(\Sigma w_k x_k\right)^r \leqslant \Sigma w_k x_k^r, \qquad \Sigma w_k = 1.$$

If we replace x_k by x_k^s, we have

$$\left(\Sigma w_k x_k^s\right)^r \leqslant \Sigma w_k x_k^{rs}.$$

Now replace rs by t:

$$\left(\Sigma w_k x_k^s\right)^{t/s} \leqslant \Sigma w_k x_k^t,$$

and finally we have

$$\left(\Sigma w_k x_k^s\right)^{1/s} \leqslant \left(\Sigma w_k x_k^t\right)^{1/t},$$

where we must have $t > s$ (since $r > 1$). When $s > t > 0$, the inequality is reversed.

Now let p_k be any positive numbers; then we may replace w_k by $p_k / \Sigma p_j$ and we find

$$\left(\frac{\Sigma p_k x_k}{\Sigma p_k}\right)^r \leqslant \frac{\Sigma p_k x_k^r}{\Sigma p_k},$$

that is,

$$\Sigma p_k x_k \leqslant \left(\Sigma p_k\right)^{1 - 1/r}.$$

Finally, let $p_k = y_k^{r/(r-1)}$, $x_k = z_k y_k^{-1/(r-1)}$; we obtain

$$\Sigma y_k z_k \leqslant \left(\Sigma y_k^{r/(r-1)}\right)^{(r-1)/r} \left(\Sigma z_k^r\right)^{1/r},$$

which is known as Hölder's inequality. The special case $r = 2$,

$$\Sigma y_k z_k \leqslant \left(\Sigma y_k^2\right)^{1/2} \left(\Sigma z_k^2\right)^{1/2},$$

is known as Cauchy's inequality.

We can deduce Minkowski's inequality (p. 23) from Cauchy's inequality. Suppose again that $\Sigma q_k = 1$; then

$$S \equiv \Sigma q_k(a_k + b_k)^2 = \Sigma q_k a_k(a_k + b_k) + \Sigma q_k b_k(a_k + b_k)$$
$$= \Sigma(q_k^{1/2}a_k)q_k^{1/2}(a_k + b_k)$$
$$+ \Sigma(q_k^{1/2}b_k)q_k^{1/2}(a_k + b_k).$$

Applying Cauchy's inequality to each sum on the right, we find

$$S \leqslant \left\{ \Sigma q_k a_k^2 \Sigma q_k(a_k + b_k)^2 \right\}^{1/2} + \left\{ \Sigma q_k b_k^2 \Sigma q_k(a_k + b_k)^2 \right\}^{1/2}$$
$$= \left\{ \Sigma q_k(a_k + b_k)^2 \right\}^{1/2} \left[\left\{ \Sigma q_k a_k^2 \right\}^{1/2} + \left\{ \Sigma q_k b_k^2 \right\}^{1/2} \right],$$

whence

$$\left\{ \Sigma q_k(a_k + b_k)^2 \right\}^{1/2} \leqslant \left\{ \Sigma q_k a_k^2 \right\}^{1/2} + \left\{ \Sigma q_k b_k^2 \right\}^{1/2}.$$

Now take $q_k = 1/n$; the preceding inequality reduces to

$$\left\{ \Sigma(a_k + b_k)^2 \right\}^{1/2} \leqslant \left\{ \Sigma a_k^2 \right\}^{1/2} + \left\{ \Sigma b_k^2 \right\}^{1/2}.$$

This could, of course, be verified more directly. The same proof, with slight modifications, using Hölder's inequality instead of Cauchy's inequality, shows that if $r > 1$,

$$\left\{ \Sigma(a_k + b_k + \cdots)^r \right\}^{1/r} \leqslant \left\{ \Sigma a_k^r \right\}^{1/r} + \left\{ \Sigma b_k^r \right\}^{1/r} + \cdots,$$

which is the general form of Minkowski's inequality.

24. Infinitely differentiable functions. We next consider functions that can be differentiated more than once, or even infinitely often. For such functions there is a generalization of the law of the mean, called Taylor's theorem with remainder. We shall not go into the motivation for considering this particular formula, and we shall not try to obtain it under the most general hypotheses possible. However, we shall obtain the formula with one of the more useful forms of the remainder.

Suppose that f is a function whose domain contains the interval $[a, x]$ and that $f^{(n)}$ exists and is continuous, or at least can be integrated to give $f^{(n-1)}$. Here $n \geqslant 1$. We start from

$$f(x) = f(a) + \int_a^x f'(t)\, dt = f(a) - \int_a^x f'(t)\, d(x - t),$$

and integrate by parts (if $n \geqslant 2$), obtaining

$$f(x) = f(a) + (x - a)f'(a) + \int_a^x (x - t)f''(t)\, dt.$$

Repeating this process, we eventually get

$$f(x) = f(a) + (x - a)f'(a) + \frac{(x - a)^2}{2} f''(a) + \cdots$$

$$+ \frac{(x - a)^{n-1}}{(n - 1)!} f^{(n-1)}(a) + R_n(x),$$

where

$$R_n(x) = \frac{1}{(n - 1)!} \int_a^x f^{(n)}(t)(x - t)^{n-1}\, dt.$$

To illustrate one of the ways in which Taylor's theorem can be used, we prove the following theorem. *Suppose that f is a function on some interval $[x_0, \infty)$, that f'' is continuous, and that $\lim_{x \to \infty} f(x) = \lim_{x \to \infty} f''(x) = 0$. Then we have $\lim_{x \to \infty} f'(x) = 0$.*

To prove this, take Taylor's theorem with remainder of order 2, and write it in the form

$$f'(a) = \frac{f(a) - f(x)}{x - a} + \frac{1}{x - a} \int_a^x (x - t) f''(t)\, dt, \quad x > a.$$

Let x_0 be so large that $|f(t)| < \epsilon$ for $t > x_0$ and $|f''(t)| < \epsilon$ for $t > x_0$. Then for $a > x_0$,

$$|f'(a)| \leqslant \frac{2\epsilon}{x - a} + \frac{\epsilon}{x - a} \int_a^x (x - t)\, dt$$

$$= \frac{2\epsilon}{x - a} + \frac{\epsilon(x - a)}{2} = \epsilon\left(\frac{2}{x - a} + \frac{x - a}{2} \right).$$

Here x is at our disposal, as long as $x > a$; take $x = a + 2$. (The reason for this particular choice is that $2/(x - a) + (x - a)/2$ is smallest when $x = a + 2$.) Then the preceding inequality reduces to $|f'(a)| \leqslant 2\epsilon$, $a > x_0$, which is to say that $f'(a) \to 0$ as $a \to \infty$.

Slight modifications in the proof will yield considerably stronger results.[53] There is no need, for example, to suppose that $f(x) \to 0$; it is enough to have $f(x)$ bounded, as long as $f''(x) \to 0$. To see this, suppose that $|f(x)| \leqslant M$ and put $\epsilon(x) = \max_{t \geqslant x}|f''(t)|$. Clearly ϵ is a nonincreasing function and $\lim_{x \to \infty}\epsilon(x) = 0$. We have

$$|f'(a)| \leqslant \frac{2M}{x - a} + \frac{1}{x - a} \int_a^x (x - t)\epsilon(t)\, dt$$

$$\leqslant \frac{2M}{x - a} + \frac{\epsilon(a)}{2}(x - a).$$

Now take x so that $x - a = \{\epsilon(a)\}^{-1/2}$. Then

$$|f'(a)| \leqslant 2M\{\epsilon(a)\}^{1/2} + \tfrac{1}{2}\{\epsilon(a)\}^{1/2},$$

and again $f'(a) \to 0$.

It is tempting to let $n \to \infty$ in Taylor's theorem and obtain an infinite series, the so-called Taylor series of $f(x)$. If $R_n(x) \to 0$ (for a particular x), the series so obtained will converge, and in fact will converge to $f(x)$. However, we must not assume that this will always happen if f has derivatives of all orders (or, as we shall say, if f is *infinitely differentiable*), although it does happen for many of the simple functions that are considered in calculus.

In the first place, the Taylor series might diverge; in the second place, it might converge, but to the wrong sum. We shall give examples of both possibilities.[54]

Even in elementary discussions, it is a commonplace that a Taylor series need not converge throughout the domain where the original function is infinitely differentiable; an example is given by $f(x) = 1/(1 + x^2)$. Here the function is infinitely differentiable on the whole of R_1, but its Taylor series (with center at 0) converges only for $|x| < 1$.

Much worse things can happen. We shall exhibit a function whose Taylor series diverges everywhere except, of course, at the point a itself. Consider the function f defined by the integral

$$f(x) = \int_0^\infty e^{-t} \cos(t^2 x)\, dt.$$

For even n we have

$$f^{(n)}(0) = \pm \int_0^\infty t^{2n} e^{-t}\, dt = \pm (2n)!,$$

and for odd n, $f^{(n)}(0) = 0$, since the derivatives of odd order of the cosine are all 0 at 0. Hence the Maclaurin series of f has as its general term

$$\pm \frac{(2n)!}{n!} x^n,$$

which does not approach zero except at $x = 0$, and therefore the series cannot converge except at $x = 0$.

It is possible to show that[55] if $\{M_k\}$ is any sequence of numbers there is an infinitely differentiable function f such that $f^{(k)}(0) = M_k$ for every k. This shows that infinitely differentiable functions whose Taylor series about 0 diverge (except at 0) must exist in great profusion.

By a more complicated construction it can be shown that there are infinitely differentiable functions whose Taylor series diverge no matter what point is taken as center.[56]

For an example of the other kind of failure of a Taylor series, consider the function f defined by

$$f(x) = e^{-1/x^2}, x \neq 0; \qquad f(0) = 0.$$

We can show that $f^{(k)}(0) = 0$ for every k, so that the Taylor series of f about 0 has all its terms 0 and thus certainly converges to the wrong sum. Clearly $f^{(k)}(x)$ has the form $R(x)e^{-1/x^2}$ for $x \neq 0$, where R is a rational function. Now $x^{-n}e^{-1/x^2} \to 0$ as $x \to 0$ for every integer n (this is equivalent to the familiar fact that $x^n e^{-x^2} \to 0$ as $x \to \infty$), and so $f^{(k)}(x) \to 0$ as $x \to 0$. This, by Exercise 21.11, implies that $f^{(k)}(0) = 0$ also. We may alternatively argue directly that

$$f^{(k)}(0) = \lim_{x \to 0} x^{-1} f^{(k-1)}(x) = 0.$$

It is not difficult to show, by direct estimates of the remainders, that the Taylor series of familiar functions such as the sine and exponential converge everywhere to the functions that gave rise to them. Similarly, the Taylor series for $(1 + x)^p$ about $x = 0$, for any real p, converges to the right value for $|x| < 1$. Since this Taylor series is just the series obtained by expanding $(1 + x)^p$ by the binomial theorem, we obtain a proof of the binomial theorem for negative or fractional exponents.

A function whose Taylor series about a converges to the function in some neighborhood of a is called *analytic* at a. As a contrast to the preceding negative examples, we prove a striking positive theorem (S. Bernstein's theorem): *If f and all its derivatives are nonnegative in an interval I, then f is analytic in the interval.* (An example is given by $f(x) = e^x$.) Suppose that $a < x < b$ and $[a, b] \subset I$. Then since $f^{(n)}(t) \geqslant 0$ we have

$$R_n(x) = \frac{1}{(n-1)!} \int_a^x f^{(n)}(t)(b-t)^{n-1}\left(\frac{x-t}{b-t}\right)^{n-1} dt$$

$$\leqslant \frac{1}{(n-1)!} \int_a^b f^{(n)}(t)(b-t)^{n-1}\left(\frac{x-t}{b-t}\right)^{n-1} dt.$$

Since $(x - t)/(b - t)$ decreases (as a function of t), it is largest at $t = a$, and we have

$$R_n(x) \leqslant \left(\frac{x-a}{b-a}\right)^{n-1} \frac{1}{(n-1)!} \int_a^b f^{(n)}(t)(b-t)^{n-1} dt$$

$$= \left(\frac{x-a}{b-a}\right)^{n-1} R_n(b).$$

But $R_n(b) \leqslant f(b)$ because all the terms of the Taylor series are nonnegative. Hence

$$R_n(x) \leqslant \left(\frac{x-a}{b-a}\right)^{n-1} f(b),$$

and since $0 < x - a < b - a$, this means that $R_n(x) \to 0$.

As a matter of fact, it is possible to replace the positivity of all derivatives by the positivity of all differences

$$\Delta_h^n f(x) = \sum_{k=0}^{n} \binom{n}{k} (-1)^k f(x + nh),$$

since it can be shown that a function with positive differences of all orders is automatically infinitely differentiable.[57] (We took the first step in §23 when we showed that a function with positive second differences has right-hand and left-hand derivatives.)

It is also true, but harder to prove, that if, on a given interval, each derivative of f has a fixed sign (possibly differing from one derivative to another), then f is analytic on the interval.[57a]

Although a nonanalytic function can have a divergent Taylor series about every point of an interval, the phenomenon just discussed, the convergence of the Taylor series to the wrong value, cannot occur throughout an interval. In fact, we can prove that *if the Taylor series of a function, about each point in an interval, has a positive radius of convergence, there must be a subinterval in which the function is analytic*. Repeated applications of this fact lead to the conclusion that (under the same hypothesis) the points about which the Taylor expansion fails can form at most a nowhere dense set.

The proof depends on a simple application of Baire's theorem. Let $\rho(a)$ denote the radius of convergence of the Taylor series of f, formed at the point a. By a familiar formula, $1/\rho(a) = \limsup_{n \to \infty} |f^{(n)}(a)/n!|^{1/n}$. Since $1/\rho(a)$ is finite for each a in the interval in question, for each a the quantity $\mu(a) = \sup_n |f^{(n)}(a)/n!|^{1/n}$ is finite. The sets E_k of points a where $\mu(a) < k$ ($k = 1, 2, \ldots$) exhaust the interval and by Baire's theorem cannot all be nowhere dense. Hence there is a subinterval in which we have

$|f^{(n)}(a)/n!|^{1/n} \leqslant k$ ($n = 0, 1, 2, \ldots$), first on a dense set, and then, by the continuity of $f^{(n)}$, throughout. In this interval f is analytic, since the last inequality shows that the remainder in the Taylor series about a approaches zero for points x such that $|x - a| < k$.

It is now natural to ask what happens when $\rho(a)$ is not only positive but bounded away from zero: $\rho(a) \geqslant \delta > 0$ for every a in an interval. It can be shown that this condition does make f analytic throughout the interval.[58]

POSTSCRIPT

The word "primer" (pronounced "primmer") was intended to suggest "a small book of elementary principles" [per *Webster's Collegiate Dictionary*, third edition]. This meaning seems to have gone out of currency. Some readers pronounce "primer" to rhyme with "climber." With this pronunciation it should mean either an undercoat of paint or something that sets off an explosion—I wonder which they have in mind.

Someone, whose name I have fortunately forgotten, once told a CUPM meeting that calculus is naturally a dull subject because nobody is doing research on it. Perhaps this book will persuade you that he was wrong on both counts.

June, 1981

NOTES

1. J. Guilloud and his collaborators have calculated a million decimal digits of π; these have not been widely distributed. For 100,000 decimal places of π, see D. Shanks and J. W. Wrench, Jr., Calculation of π to 100,000 decimals, *Math. of Computation* 16 (1962), 76–99.   [p. 14]

1a. For alternative proofs see M. Reichbach, Une simple démonstration du théorème de Cantor-Bernstein, *Colloquium Math.* 3 (1955), 163; M. S. Hellmann, A short proof of an equivalent form of the Schroeder-Bernstein theorem, *Amer. Math. Monthly* 68 (1961), 770; M. F. Smiley, *Algebra of matrices*, Allyn and Bacon, Boston, 1965, pp. 235–236; R. H. Cox, A proof of the Schroeder-Bernstein theorem, *Amer. Math. Monthly* 75 (1968), 508.   [p. 18]

2. See, for example, C. Kuratowski, *Topologie*, vol. 2, *Monografie Matematyczne*, vol. 21, 2nd ed., Warsaw, 1952, p. 85.
                                                            [p. 32]

2a. See, for example, J. Cobb and W. Voxman, Dispersion points and fixed points, *Amer. Math. Monthly* 87 (1980), 278–281.   [p. 41]

2b. See John Sack, *Report from practically nowhere*, New York, 1959, p. 23.   [p. 45]

3. Proof suggested by W. C. Fox and R. R. Goldberg.   [p. 45]

4. See, for example, M. E. Munroe, *Introduction to measure and integration*, Addison-Wesley, Cambridge, Mass., 1953, p. 30.
                                                            [p. 49]

5. Any method, other than convergence, for attaching a sum to an infinite series is called a method of summability. See O. Szász, *Introduction to the theory of divergent series*, Hafner, New

York, 1948; G. H. Hardy, *Divergent series*, Oxford, 1949; K. Zeller, *Theorie der Limitierungsverfahren*, *Ergebnisse der Mathematik*, new series, no. 15, Springer, Berlin-Göttingen-Heidelberg, 1958.   [p. 53]

6. See, for example, H. Hahn, *Reelle Funktionen*, Akademische Verlagsgesellschaft, Leipzig, 1932, p. 115.   [p. 54]

7. E. Corominas and F. Sunyer Balaguer, Condiciones para que una función infinitamente derivable sea un polinomio, *Revista Mat. Hisp.-Amer.* (4) 14 (1954), 26–43. Several extensions of the theorem are given in this paper.   [p. 65]

8. C. E. Weil (On nowhere monotone functions, *Proc. Amer. Math. Soc.* 56 (1976), 388–389) gave a short proof of the existence of such functions via Baire's theorem, and a number of references to other constructions. See also, A. Denjoy, Sur les fonctions dérivées sommables, *Bull. Soc. Math. France* 43 (1915), 161–248, especially pp. 228 ff.   [p. 68]

9. S. Banach, Uber die Baire'sche Kategorie gewisser Funktionenmengen, *Studia Math.* 3 (1931), 174–180.   [p. 68]

10. The simplest example was constructed by A. P. Morse, A continuous function with no unilateral derivatives, *Trans. Amer. Math. Soc.* 44 (1938), 496–507.   [p. 68]

11. S. Saks, On the functions of Besicovitch in the space of continuous functions, *Fund. Math.* 19 (1932), 211–219.   [p. 68]

11a. This is a special case of Sierpiński's theorem that a compact connected set containing more than one point is never a countable union of disjoint closed sets; for some references see *Amer. Math. Monthly* 84 (1977), 828.   [p. 69]

11b. See also G. J. Minty, On the notion of "function," *Amer. Math. Monthly* 78 (1971), 188–189.   [p. 75]

12. See especially J. B. Rosser, *Logic for mathematicians*, McGraw-Hill, New York, 1953, pp. 306 ff.; K. Menger, *Calculus, a modern approach*, Ginn, Boston, 1955.   [p. 77]

13. G. H. Hardy, A formula for the prime factors of any number, *Messenger of Math.* 35 (1906), 145–146.   [p. 78]

14. W. Sierpiński, Sur un exemple effectif d'une fonction non représentable analytiquement, *Fund. Math.* 5 (1924), 87–91.

      [p. 78]

15. H. Lebesgue, *Leçons sur l'intégration et la recherche des fonctions primitives*, 2nd ed., Gauthier-Villars, Paris, 1928, p. 97.
[p. 79]

15a. H. Fast, Une remarque sur la propriété de Weierstrass, *Colloq. Math.* 7 (1959), 75–77. The sum of a nonconstant continuous function and a function with the intermediate value property can fail to have the intermediate value property. For more information about properties of this kind, see A. M. Bruckner and J. Ceder, On the sum of Darboux functions, *Proc. Amer. Math. Soc.* 51 (1975), 97–102.   [p. 80]

15b. H. Blumberg, New properties of all real functions, *Trans. Amer. Math. Soc.* 24 (1922), 113–128.   [p. 83]

15c. W. Sierpiński and A. Zygmund, Sur une fonction qui est discontinue sur tout ensemble de puissance du continu, *Fund. Math.* 4 (1923), 316–318.   [p. 83]

15d. The last statement also follows from the definition of continuity on p. 83, since the property in question requires the inverse image of every open interval to be an open interval (and hence the inverse image of every open set to be an open set). For a more thorough treatment see J. B. Diaz, Discussion and extension of a theorem of Tricomi concerning functions which assume all intermediate values, *J. Math. Mech.* 18 (1968/69), 617–628; also the review of this paper in *Math. Reviews* 39 #370. For a localized version of the property just discussed see E. W. Chittenden, Note on functions which approach a limit at every point of an interval, *Amer. Math. Monthly* 25 (1918), 249–250.   [p. 88]

15e. More generally there is no continuous transformation of an interval such that each image point has exactly two inverses. See O. G. Harrold, The non-existence of a certain type of continuous transformation, *Duke Math. J.* 5 (1939), 789–793; and for extensions, J. H. Roberts, Two-to-one transformations, *Duke Math. J.* 6 (1940), 256–262; P. Civin, Two-to-one mappings of manifolds, *Duke Math. J.* 10 (1943), 49–57.

There is a substantial amount of literature on related topics. See, for example, papers by A. V. Černavskiĭ, J. Mioduszewski, W. R. Utz, and B. R. Wenner, which can be located through

Mathematical Reviews; and Problems E 1094 and E 1715, *Amer. Math. Monthly* (1954, 425; 1965, 784).   [p. 89]

15f. D. C. Gillespie, A property of continuity, *Bull. Amer. Math. Soc.* 28 (1922), 245–250.   [p. 89]

15g. Gillespie (*loc. cit.*) gives a formula for such a function: $f(x) = \pi x + x^2 \sin(\pi/x)$, $0 < x \leqslant 1$.   [p. 89]

15h. J. B. Diaz and F. T. Metcalf, A continuous periodic function has every chord twice, *Amer. Math. Monthly* 74 (1967), 833–835, give a different proof. For extensions to almost periodic and more general functions, see J. C. Oxtoby, Horizontal chord theorems, *Amer. Math. Monthly* 79 (1972), 468–475.   [p. 92]

16. Although T. M. Flett (*Bull. Inst. Math. Appl.* 11 (1975), 34) has discovered that the positive part of the universal chord theorem was proved by A. M. Ampère in 1806 (see *Math. Rev.* 56 #5805)), the modern history of the theorem begins with P. Lévy, Sur une généralisation du théorème de Rolle, *C. R. Acad. Sci. Paris* 198 (1934), 424–425; it has been repeatedly rediscovered. For extensions, see H. Hopf, Über die Sehnen ebener Kontinuen und die Schliefen geschlossener Wege, *Comment. Math. Helv.* 9 (1937), 303–319. For a discussion of the possible lengths of horizontal chords for a given function see Hopf's paper; also Oxtoby's paper cited in note 15h, and R. J. Levit, The finite difference extension of Rolle's theorem, *Amer. Math. Monthly* 70 (1963), 26–30. For further information and some applications see J. T. Rosenbaum, Some consequences of the universal chord theorem, *Amer. Math. Monthly* 78 (1971), 509–513. Rosenbaum also gave a picturesque interpretation of the theorem: How to build a picnic table for use on a mountain range of known period, *Notices Amer. Math. Soc.* 16 (1969), 94.

Another application was shown to me by J. C. Oxtoby: if f is continuous and $f(x + y) = g(f(x), y)$ for all x and y, then f is either strictly monotonic or constant. (This is problem E2246, *Amer. Math. Monthly* 78 (1971), 676–677.) If f is not strictly monotonic, $f(x_0 + h) = f(x_0)$ for some x_0 and some $h > 0$. Then for all x

$$f(x + h) = f((x_0 + h) + (x - x_0)) = g(f(x_0 + h), x - x_0)$$

$$= g(f(x_0), x - x_0) = f(x_0 + (x - x_0)) = f(x).$$

Since f has arbitrarily short horizontal chords (by the universal chord theorem), this formula shows that f has arbitrarily short periods, and so is constant. Oxtoby remarks that conversely if f is continuous and either strictly monotonic or constant, there is a function g such that $f(x + y) = g(f(x), y)$.   [p. 92]

17. Hopf, note 16.   [p. 93]

17a. Suggested by J. D. Memory, Kinematics problem for joggers, *Amer. J. Physics* 41 (1973), 1205–1206.   [p. 95]

18. For an elementary and detailed exposition of this and related theorems, see A. W. Tucker, Some topological properties of disk and sphere, *Proceedings of the First Canadian Mathematical Congress 1945*, University of Toronto Press, 1946, pp. 285–309.   [p. 95]

18a. See W. G. Chinn and N. E. Steenrod, *First concepts of topology* (New Mathematical Library, no. 18), Random House, New York, 1966, p. 65.   [p. 96]

19. J. G. Brennan, A property of a plane convex region, *Math. Gaz.* 42 (1958), 301–302; A. C. Zitronenbaum, Bisecting an area and its boundary, *Math. Gaz.* 43 (1959), 130–131.   [p. 96]

20. For a proof, references, and extensions, see A. H. Stone and J. W. Tukey, Generalized "sandwich" theorems, *Duke Math. J.* 9 (1942), 356–359; Chinn and Steenrod, book cited in 18a, p. 120.   [p. 97]

20a. For an interesting analysis of the concept of continuity in terms of limits, see K. P. Williams, Note on continuous functions, *Amer. Math. Monthly* 25 (1918), 246–249.   [p. 99]

21. For this and the following theorem see G. Pólya and G. Szegő, *Aufgaben und Lehrsätze aus der Analysis*, Springer, Berlin, 1925, vol. 1, pp. 63, 225, problems II 126, 127.   [p. 104]

22. For analyses of the notion of continuous curve from different points of view, see G. T. Whyburn, What is a curve?, *Amer. Math. Monthly* 49 (1942), 493–497; J. W. T. Youngs, Curves and surfaces, *ibid.* 51 (1944), 1–11.   [p. 106]

23. W. Hurewicz, Über dimensionserhöhender stetige Abbildungen, *J. Reine Angew. Math.* 169 (1933), 71–78.   [p. 106]

24. I. J. Schoenberg, On the Peano curve of Lebesgue, *Bull. Amer. Math. Soc.* 44 (1938), 519.   [p. 106]

25. For extensions, see L. Lorch, Derivatives of infinite order, *Pacific J. Math.* 3 (1953), 773–778.   [p. 110]

26. Pólya and Szegő, book cited in note 21, vol. 1, pp. 30, 185, problem I 165.　　　　　　　　　　　　　　　　　　　　　　[p. 112]

26a. See, for example, T. M. Apostol, *Mathematical Analysis*, Addison-Wesley, Reading, Mass., 1957, p. 458.　　　　　　[p. 112]

27. See, for example, C. de la Vallée Poussin, *Intégrale de Lebesgue, fonctions d'ensemble, classes de Baire*, 2nd ed., Gauthier-Villars, Paris, 1934, pp. 127 ff.　　　　　　　　　[p. 115]

27a. For a simple proof see F. W. Carroll, Separately continuous functions are Baire functions, *Amer. Math. Monthly* 78 (1971), 175.　　　　　　　　　　　　　　　　　　[p. 117]

28. The theorem that a continuous function can be uniformly approximated by polynomials is known as the Weierstrass approximation theorem. The proof given here was invented by E. Landau.　　　　　　　　　　　　　　　　　　　　　　[p. 121]

29. What is needed is the existence of a Hamel basis for the real numbers: see H. Hahn and A. Rosenthal, *Set functions*, University of New Mexico Press, Albuquerque, 1948, pp. 100 ff. A discontinuous linear function is constructed in G. H. Hardy, J. E. Littlewood, and G. Pólya, *Inequalities*, Cambridge University Press, 1934, p. 96. Hamel's original paper is cited in note 32. See also G. S. Young, The linear functional equation, *Amer. Math. Monthly* 65 (1958), 37–38. Additional history and references are given by J. W. Green and W. Gustin, Quasi-convex sets, *Canadian J. Math.* 2 (1950), 489–507.　　　　　　　　[p. 125]

30. For this proof, and extensions, see H. Kestelman, On the functional equation $f(x + y) = f(x) + f(y)$, *Fund. Math.* 34 (1947), 144–147.　　　　　　　　　　　　　　　　　　[p. 126]

31. A. Zygmund, *Trigonometrical series, Monografje Matematyczne*, vol. 5, Warsaw-Lwów, 1935, pp. 133–134; *Trigonometric series*, vol. 1, Cambridge University Press, 1959, p. 235. The property was discovered by H. Steinhaus. For a simple arithmetical proof see J. F. Randolph, Distances between points of the Cantor set, *Amer. Math. Monthly* 47 (1940), 549–551. For further interesting related properties of the Cantor set and similar sets, see some recent papers by T. Šalát (*Math. Reviews* 24 #A2538) and N. C. Bose Majumder (*Math. Reviews* 22 #2971; 24 #A1706, A2537, A3444; 29 #5215, 5732, 5733), *Amer. Math.*

Monthly 72 (1965), 725–729 (*Math. Reviews* 32 #1295). See also J. M. Brown and K. W. Lee, The distance set of $C_\lambda \times C_\lambda$, *J. London Math. Soc.* (2) 15 (1977), 551–556.   [p. 127]

32. G. Hamel, Eine Basis aller Zahlen und die unstetigen Lösungen der Funktionalgleichung $f(x + y) = f(x) + f(y)$, *Math. Ann.* 60 (1905), 459–462.   [p. 127]

33. M. Plancherel and G. Pólya, Sur les valeurs moyennes des fonctions réelles définies pour toutes les valeurs de la variable, *Comment. Math. Helv.* 3 (1931), 114–121.   [p. 129]

33a. See R. P. Agnew, Limits of integrals, *Duke Math. J.* 9 (1942), 10–19; Mean values and Frullani integrals, *Proc. Amer. Math. Soc.* 2 (1951), 237–241; Frullani integrals and variants of the Egoroff theorem on essentially uniform convergence, *Acad. Serbe Sci. Publ. Inst. Math.* 6 (1954), 12–16.   [p. 130]

33b. For an extensive survey of theorems about derivatives, see A. M. Bruckner, *Differentiation of Real Functions*, Lecture Notes in Mathematics 659, Springer, 1978.   [p. 130]

33c. Mikolás, note 34. A similar condition that implies continuous differentiability is given by G. P. Weller, A condition implying differentiability, *Delta* 4, no. 2 (1974), 59–64.   [p. 132]

34. M. Mikolás, Construction des familles de fonctions partout continues non dérivables, *Acta Sci. Math. Szeged* 17 (1956), 49–62. For other simple constructions, see J. McCarthy, An everywhere continuous nowhere differentiable function, *Amer. Math. Monthly* 60 (1953), 709; T. H. Hildebrandt, A simple continuous function with a finite derivative at no point, *Amer. Math. Monthly* 40 (1933), 547–548.   [p. 132]

34a. U. Dini, Grundlagen für eine Theorie der Functionen einer veränderlichen reellen Grösse, translated from the Italian edition of 1878, Leipzig, 1892, pp. 279–280. Note that a function can be "increasing on the right at a point" without necessarily being increasing in any right-hand interval.   [p. 134]

35. W. Sierpiński: see S. Saks, Théorie de l'intégrale, *Monografje Matematyczne*, vol. 2, Warsaw, 1933, pp. 167–168. [p. 134]

36. S. Saks, book cited in note 35, p. 177. See also D. E. Varberg, On absolutely continuous functions, *Amer. Math. Monthly* 72 (1965), 831–841; E. P. Woodruff, Derivates of a

function whose image is of Lebesgue measure zero, *Notices Amer. Math. Soc.* 16 (1969), 666–667. For an elementary proof of a more elementary result, see Pólya and Szegő, book cited in note 21, vol. 1, pp. 63, 225, problem II 125.   [p. 135]

36a. One could claim that this dates back to Galileo, who wrote in 1638 (Discorsi e dimostrazioni matematiche intorno a due nuove scienze, *Opera*, 1898, vol. 8, p. 283), "Plato perhaps had the idea that a moving object cannot pass from rest to a positive velocity without passing through all lower velocities." It seems likely that Galileo would have thought of velocity as being continuous, so he may merely have noticed the intermediate value property of continuous functions. The tentative attribution to Plato seems to be fanciful.   [p. 136]

36b. A similar diagram once appeared on a Graduate Record Examination with directions something like "Which of the following 5 theorems does the picture make you think of?"

[p. 136]

36c. See A. K. Aziz and J. B. Diaz, On Pompeiu's proof of the mean-value theorem of the differential calculus of real-valued functions, *Contributions to Differential Equations*, vol. 1, pp. 467–481 (1963). See also H. Samelson, On Rolle's theorem, *Amer. Math. Monthly* 86 (1979), 486.   [p. 137]

36d. Cf. J. Dieudonné: *Foundations of Modern Analysis*, Academic Press, 1960, pp. 153–155.   [p. 137]

37. E. W. Hobson, *The theory of functions of a real variable and the theory of Fourier's series*, vol. 1, 3rd ed., Cambridge University Press, 1927, p. 363.   [p. 138]

38. T. M. Flett, A mean value theorem, *Math. Gaz.* 42 (1958), 38–39. See also S. G. Wayment, An integral mean value theorem, *Math. Gaz.* 54 (1970), 300–301, and references given there; J. B. Diaz and R. Výborný, On some mean value theorems of the differential calculus, *Bull. Austral. Math. Soc.* 5 (1971), 227–238; S. Reich, Problem 5810, *Amer. Math. Monthly* 78 (1971), 798.

[p. 140]

39. A special case is considered by L. J. Paige, A note on indeterminate forms, *Amer. Math. Monthly* 61 (1954), 189–190.

[p. 140]

39a. See A. P. Morse, Dini derivates of continuous functions, *Proc. Amer. Math. Soc.* 5 (1954), 126–130.   [p. 141]

40. For references see P. Erdős, Some remarks on set theory, *Ann. of Math.* 44 (1943), 643–646 (p. 646).   [p. 144]

41. A. N. Singh, On infinite derivatives, *Fund. Math.* 33 (1945), 106–107.   [p. 144]

41a. F. M. Filipczak, On the derivative of a discontinuous function, *Colloq. Math.* 13 (1964), 73–79; K. M. Garg, On bilateral derivatives and the derivative, *Trans. Amer. Math. Soc.* 210 (1975), 295–329 (Proposition 3.9 and Corollary 5.2); R. P. Boas and G. T. Cargo, Level sets of derivatives, *Pacific J. Math.* 83 (1979), 37–44.   [p. 145]

42. This seems first to have been noticed explicitly by M. K. Fort, Jr. in 1951: A theorem concerning functions discontinuous on a dense set, *Amer. Math. Monthly* 58 (1951), 408–410. It was found independently by S. Četković, Un théorème de la théorie des fonctions, *C. R. Acad. Sci. Paris* 245 (1957), 1692–1694; and by S. Markus [Marcus], Points of discontinuity and points of differentiability (in Russian), *Rev. Math. Pures Appl.* (2) (1957), 471–474. However, it is an easy consequence of an old theorem of W. H. Young (On the infinite derivates of a function of a single variable, *Arkiv för Matematik, Astronomi och Fysik* 1 (1903), 201–204). Young's theorem states that the set of points at which at least one Dini derivate is infinite is a countable intersection of open sets (hence, if it is dense, its complement is of first category). That Fort's theorem is a corollary of Young's was noticed by K. M. Garg, On the derivability of functions discontinuous at a dense set, *Rev. Math. Pures Appl.* 7 (1962), 175–179, and independently by Cargo (see Boas and Cargo, op. cit.; that paper also contains a particularly simple proof of Young's theorem).

The proof of Fort's theorem in the text was suggested by A. C. Segal. The longer proof given in the first edition of this book had a misprint: the second $>$ on p. 127, line 6, should have been $<$. See also *Amer. Math. Monthly* 78 (1971), p. 1106. Another simple proof is implicit in more general results obtained by E. M. Beesley, A. P. Morse, and D. C. Pfaff, Lipschitzian points, *Amer. Math. Monthly* 79 (1972), 603–608.   [p. 146]

43. For a simple exposition see F. Riesz and B. Sz.-Nagy, *Functional analysis*, Ungar, New York, 1955, pp. 17 ff.   [p. 147]

43a. See A. M. Bruckner, Creating differentiability and destroying derivatives, *Amer. Math. Monthly* 85 (1978), 554–562.
[p. 148]

44. J. S. Lipiński, Sur la dérivée d'une fonction de sauts, *Colloq. Math.* 4 (1957), 197–205; L. A. Rubel, Differentiability of monotonic functions, *Colloq. Math.* 10 (1963), 277–279.   [p. 150]

45. This simple construction was given by G. Freilich, *Amer. Math. Monthly* 80 (1973), 918–919.   [p. 153]

46. The construction given in the first edition of this book was fallacious. The present construction is modeled on that given by S. Saks (*Theory of the integral, Monografie Matematyczne*, vol. 7, Warsaw-Lwów, 1937, pp. 205–206) for the more general situation when the derivative is infinite on an arbitrary closed set of measure 0. Some interesting results in this connection are: *If g is continuous and has a (finite or infinite) derivative except on a countable set, this derivative being nonnegative almost everywhere, then g is nondecreasing* (G. Goldowsky; see Saks, loc. cit.); *if E is a countable set which is the intersection of countably many open sets, there is a (not necessarily continuous) function whose derivative is* $+\infty$ *on E and* 0 *outside E* (G. Piranian, The derivative of a monotonic discontinuous function, *Proc. Amer. Math. Soc.* 16 (1965), 243–244).   [p. 154]

47. The exposition follows Riesz and Sz.-Nagy, book cited in note 43, with some simplifications given by H. Kestelman, *Modern theories of integration*, Oxford, 1937, pp. 199 ff. The theorem was originally proved by Lebesgue as a deduction from his entire theory of integration. A further simplification has been given by L. A. Rubel (paper cited in note 44), and a very short proof along different lines has been given by D. G. Austin, A geometric proof of the Lebesgue differentiation theorem, *Proc. Amer. Math. Soc.* 16 (1965), 220–221.   [p. 155]

48. This proof and that of the next theorem follow Riesz and Sz.-Nagy, book cited in note 43. "Fubini's theorem" in the theory of integration is a different theorem.   [p. 160]

49. P. Erdős, Some remarks on the measurability of certain

sets, *Bull. Amer. Math. Soc.* 51 (1945), 728–731 (proof by T. Radó).   [p. 163]

50. The most general result is given by A. Ostrowski, Zur Theorie der konvexen Funktionen, *Comment. Math. Helv.* 1 (1929), 157–159.   [p. 164]

51. The exposition follows R. P. Boas and D. V. Widder, Functions with positive differences, *Duke Math. J.* 7 (1940), 496–503.   [p. 165]

51a. I learned of this application from M. Klamkin. A forthcoming book by I. Niven will discuss many similar problems.   [p. 172]

52. J. R. Nurcombe, A necessary condition for convergence, *Math. Gaz.* 59 (1975), 113–114.   [p. 173]

52a. For the inequalities discussed here see Hardy, Littlewood, and Pólya, *Inequalities*, Cambridge University Press, 1934, chapters 2 and 3. See also E. F. Beckenbach and R. Bellman, *Inequalities*, Springer-Verlag, Berlin-Heidelberg-New York, 1961, 1965; D. S. Mitrinović and P. M. Vasić, *Analytic Inequalities*, Springer-Verlag, 1970.   [p. 174]

53. For references and generalizations, see Boas, Asymptotic relations for derivatives, *Duke Math. J.* 3 (1937), 637–646.
[p. 178]

54. For literature see H. Salzmann and K. Zeller, Singularitäten unendlich oft differenzierbarer Funktionen, *Math. Z.* 62 (1955), 354–367.   [p. 179]

55. E. Borel. See A. Rosenthal, On functions with infinitely many derivatives, *Proc. Amer. Math. Soc.* 4 (1953), 600–602; Salzmann and Zeller, paper cited in note 54; H. Mirkil, Differentiable functions, formal power series, and moments, *Proc. Amer. Math. Soc.* 7 (1956), 650–652.   [p. 180]

56. For a proof by Baire's theorem see D. Morgenstern, Unendlich oft differenzierbare nicht-analytische Funktionen, *Math. Nachr.* 12 (1954), 74; for a more explicit construction see H. Cartan, *Sur les classes de fonctions définies par des inégalités portant sur leurs dérivées successives*, *Actualités Scientifiques et Industrielles*, no. 867 (1940), pp. 20–22.   [p. 180]

57. S. Bernstein. See Boas and Widder, paper cited in note 51.
[p. 182]

57a. For many years there was no elementary proof, but one was found recently: J. A. M. McHugh, A proof of Bernstein's theorem on regularly monotonic functions, *Proc. Amer. Math. Soc.* 47 (1975), 358–360. [p. 182]

58. See Salzmann and Zeller, paper cited in note 54. [p. 183]

59. Cf. W. F. Osgood, Beweis der Existenz einer Lösung der Differentialgleichung $dy/dx = f(x, y)$ ohne Hinzunahme der Cauchy-Lipschitz'schen Bedingung, *Monatsh. Math. Phys.* 9 (1898), 331–345; p. 344. [p. 219]

ANSWERS TO EXERCISES

1.1. Restatement of the definition.

1.2. (a) Every letter is either a consonant or a vowel and all the vowels occur in "real functions."

(b) $C(E)$ consists of all the vowels.

(c) $C(F)$ consists only of consonants (in fact, not all of them).

(d) $F \cap E = \{r, l, f, n, c, t, s\}$ contains no vowels.

1.2a. Not very practical. (Repeat the preceding discussion with "bibliography" replacing "set.")

2.1. (a) All numbers greater than or equal to 1; all nonpositive numbers; 1; 0.

(b), (c), (d), (e): The same as (a).

(f) All nonnegative numbers; all nonpositive numbers; 0; 0.

(g), (h): $+\infty$.

(i) All numbers greater than or equal to $\pi/6$; $\pi/6$; $5\pi/6$.

2.2. Every nonempty set E that is bounded below has a greatest lower bound, denoted by $\inf E$ (for infimum), with the properties that every $x \in E$ satisfies $x \geqslant \inf E$, and that if $A > \inf E$ there is at least one $x \in E$ such that $x < A$. If the least upper bound property is assumed and E is a set which is bounded below, let F consist of all numbers x such that $-x \in E$. Since $x > M$ means $-x < -M$, the set F is bounded above and so has a least upper bound B. Then $-B$ is the greatest lower bound of E. For, if $x \in E$, $-x \leqslant B$, so $x \geqslant -B$; if $A > -B$, $-A < B$, there is an $x \in E$ with $-x > -A$, so $x < A$.

2.3. $+\infty + (-\infty)$ would be $(a/0) + (b/0)$ where $a > 0$, $b < 0$. To preserve the rules of arithmetic this would have to equal $(a + b)/0$, which can be $+\infty$, $-\infty$, or undefined according as $a + b > 0$, $a + b < 0$, or $a + b = 0$. If $0 \cdot (+\infty)$

$= x$, then $+\infty = x/0$ and x can be any positive number. It is also impossible to attach a meaning to $(+\infty)/(+\infty)$.

2.4. If E contains the point x, every upper bound of E is at least x and every lower bound of E is at most x. The same statement is true with y in place of x if E also contains a point y. If $y > x$, then, we have in particular $\sup E \geqslant y > x \geqslant \inf E$. Similarly if $y < x$.

3.1. Let the sets be E and F, and discard from F any elements it may have in common with E. If the reduced set (F_1, say) is finite, count F_1 and then E. If not, use the odd positive integers to label F_1 and the even positive integers to label E.

3.2. The elements of the kth set, E_k, may be denoted by $e_{k,1}, e_{k,2}, \ldots$. Associate $e_{k,n}$ with the lattice point (k, n).

3.2a. Each disk contains a point with two rational coordinates that is not in any other disk; enumerate S by enumerating these points.

3.3. The linear polynomials $ax + b$ with integral coefficients are in one-to-one correspondence with the lattice points (a, b); the quadratic polynomials $ax^2 + bx + c$, with the three-dimensional lattice points (a, b, c); and so on.

3.4. If the real numbers x in (a, b) were countable, the real numbers $x - a$ in $(0, b - a)$ would be countable and so would the real numbers $(x - a)/(b - a)$ in $(0, 1)$.

3.5. The construction of the text applies word for word, since the number constructed contains no 3's.

3.5a. The suggested procedure fails to count the real numbers that have infinitely many nonzero digits (i.e., most of them).

3.6. If E with x_1 removed is finite, it is still finite after adjoining the one point x_1. If the process terminated at some stage, say after removing $x_1, \ldots, x_k$, we should have a finite set left, and this set would still be finite after adjoining k points.

3.7. Let F be E with x_0 deleted. Select a countably infinite subset $\{x_1, x_2, \ldots\}$ of F. Let points of E, other than $x_0, x_1, \ldots$, correspond to themselves (as points of F); let x_0 correspond to x_1, x_1 to x_2, etc.

3.8. The correspondence $x \leftrightarrow \tan x$ establishes a one-to-one correspondence between the real numbers between $-\pi/2$ and $\pi/2$ and the set of all real numbers. Geometrical arguments can also be used.

3.9. If "the set of all sets" is a set S, the aggregate of subsets of S is a set T that cannot be put into one-to-one correspondence with any subset of S. On the other hand, the subsets of S are sets and so elements of S, and their aggregate T is therefore in one-to-one correspondence with a subset of S, namely T itself. Contradiction.

3.9a. If the functions could be put into one-to-one correspondence with the real numbers, each f could be labeled as f_x, where x is the real number that corresponds to the function f. Define a function g by $g(x) = f_x(x) + 1$. What x could g correspond to?

3.10. Use the Schroeder-Bernstein theorem. On the one hand, to each real number we can assign a sequence of real numbers, namely, the sequence of digits of the decimal that defines the real number. On the other hand, if we have a sequence of real numbers, we can write out their decimal expansions, one below another, and traverse the resulting array diagonally to obtain the decimal expansion of a real number. Different sequences of real numbers generate different real numbers in this way, since the decimal we obtain cannot terminate.

4.0. (a) No. $d(x, y)$ can be 0 without having $x = y$. In fact, $d((0, a), (0, b)) = 0$.

(b) No; $d(x, y)$ is not always equal to $d(y, x)$; indeed, $d(x, y)$ is not always well-defined.

(c) No; $d(x, y)$ is not symmetric.

(d) Yes. Properties (1) and (2) are obvious. Write $x = e^u$, $y = e^v$, $z = e^w$. Write the triangle inequality in terms of u, v, w.

(e) No: $d(10^{-11}, 0) = 0$.

4.1. Properties (1) and (2) of both new metrics are obvious. For property (3), let the points denoted originally by x, y, z be $(x_1, y_1), (x_2, y_2), (x_3, y_3)$. We have to verify

$$|x_1 - x_3| + |y_1 - y_3| \leqslant |x_1 - x_2| + |y_1 - y_2| \\ + |x_2 - x_3| + |y_2 - y_3|$$

and

$$\max(|x_1 - x_3|, |y_1 - y_3|) \leqslant \max(|x_1 - x_2|, |y_1 - y_2|) \\ + \max(|x_2 - x_3|, |y_2 - y_3|).$$

The first follows from $|x_1 - x_3| \leqslant |x_1 - x_2| + |x_2 - x_3|$ (and the same with x replaced by y). The second follows from the same inequalities, since

$$\left.\begin{array}{c} |x_1 - x_3| \\ |y_1 - y_3| \end{array}\right\} \leqslant \left\{\begin{array}{c} |x_1 - x_2| + |x_2 - x_3| \\ |y_1 - y_2| + |y_2 - y_3| \end{array}\right\}$$

$$\leqslant \max(|x_1 - x_2|, |y_1 - y_2|)$$
$$+ \max(|x_2 - x_3|, |y_2 - y_3|).$$

In the triangle with vertices $(0, 0)$, $(0, 1)$, and $(1, 0)$, the lengths of the sides are 1, 1, and 2 in the first new metric. In the triangle with vertices $(0, 0)$, $(1, -1)$, and $(1, 1)$, the lengths of the sides are 1, 1, and 2 in the second new metric. The locus $d(x, 0) = 1$ in the first new metric consists of points (x, y) such that $|x| + |y| = 1$, that is, it is the square with vertices $(0, \pm 1), (\pm 1, 0)$. In the second metric the locus consists of points (x, y) such that $|x| = 1$ and $|y| \leqslant 1$, or $|x| \leqslant 1$ and $|y| = 1$; this is the square with vertices at the four points $(\pm 1, \pm 1)$.

4.2. All the requirements for a metric space are still fulfilled.

4.3. $d(x, x) = 0; d(y, x) = d(x, y) = 1 > 0$ if $x \neq y$; if $x = z$, $0 = d(x, z) \leqslant d(x, y) + d(y, z)$; if $x \neq z$, $1 = d(x, z) \leqslant d(x, y) + d(y, z)$, since $x = y = z$ is excluded.

5.1. A neighborhood of the point x in C consists of all continuous functions y such that $|y(t) - x(t)| < r$ for all t in $[0, 1]$.

5.2. If $p = (m, n)$ is a point of the space, a neighborhood of p consists of all points with two integral coordinates at (ordinary) distance less than r from p. Each neighborhood contains only a finite number of points, and if $r < 1$, it contains only its center.

5.2.a. (i) Empty interior; all points are boundary points.

(ii) The set is its own interior. Its boundary is $1, \frac{1}{2}, \frac{1}{3}, \ldots,$ and 0.

(iii) The boundary consists of line segments of length 1 standing on the boundary points in (ii).

(iv) Empty interior; the boundary is the whole set.

5.3. Neighborhoods of radius less than 1 contain only their centers (Exercise 5.2). Hence sufficiently small neighborhoods of any point of any nonempty set E never contain points of $C(E)$, so the boundary of E is empty.

5.4. The definition is symmetric in E and $C(E)$.

5.5. Let x be a boundary point of B and let N be a neighborhood of x. Then N contains at least one point y of B; N also contains a neighborhood of y, and this neighborhood contains a point of E and a point of $C(E)$. Thus x is a boundary point of E, that is, $x \in B$. If E consists of the rational points of R_1, its boundary is all of R_1, so that the boundary of the boundary of E is empty.

5.6. (a) The boundary of N consists of the points at distance r from x.

(b) The boundary of N may be empty (Exercise 5.3), but if it has points they must be at distance r from x. For, if a point y of N is at distance less than r from x, so is a small neighborhood of y, so that y is an interior point. Similarly, a point y at distance greater than r from x is an interior point of $C(E)$.

5.7. If $a < x < b$, a neighborhood of x is an interval $(x - h, x + h)$, and if $h < \min(x - a, b - x)$, this interval is inside (a, b). Hence x is an interior point.

The boundary points of $[a, b]$ are therefore the points a and b, and these are in $[a, b]$, so $[a, b]$ is closed.

5.8. (a, b) is neither open nor closed in R_2, since it has an empty interior but fails to contain the points a and b of its boundary. But $[a, b]$ is closed in R_2 since now all its points are boundary points.

5.9. 0 is not an interior point; 1 is a boundary point that is not in the set.

5.10. All the points of the whole space are interior points, so the space is open. The boundary of the whole space is empty, and therefore contained in the space. Hence the whole space is both open and closed. The empty set contains its (empty) interior and its (empty) boundary.

5.11. The intervals $(n, n + \frac{1}{2})$ are sets of the required kind; so are unions of such sets.

5.12. Neither: its boundary is all of R_1 and its interior is empty.

5.13. Any neighborhood is open. A neighborhood such as the set of points of the space at distance less that $\sqrt{2}$ from 0 is also closed, since its boundary is empty.

5.14. If E is any set in this space, neighborhoods of radius less than 1 of points of E belong to E, so that E is open. The boundary of E is empty (Exercise 5.3), and therefore contained in E; hence E is closed.

5.15. If $x \subset G \subset E$ where G is open, x is an interior point of G and therefore an interior point of E, since a neighborhood of x consisting exclusively of points of G also consists exclusively of points of E.

5.16. If E is open and $x \in E$, some neighborhood of x contains only points of E, so not all neighborhoods of x can contain both points of E and points of $C(E)$. Hence E contains none of its boundary points. Conversely, suppose that E contains none of its boundary points and let $x \in E$. Then x is not a boundary point, whence some neighborhood of x fails to contain any points of $C(E)$ and so consists exclusively of points of E. Hence E is open.

5.17. E is open if and only if it contains none of its boundary points, therefore if and only if all the boundary points of E belong to $C(E)$; that is, if and only if $C(E)$ contains all the boundary points of $C(E)$ (since the boundary of E is the boundary of $C(E)$); that is, if and only if $C(E)$ is closed.

5.18. Reformulation of Exercise 5.17: interchange E and $C(E)$.

5.19. Let E contain all its boundary points and let x be a limit point of E. Then either $x \in E$ or $x \in C(E)$. In the second case, every neighborhood of x contains points of E (since x is a limit point of E) and a point of $C(E)$ (x itself). Therefore x is a boundary point of E if not in E. Hence if E contains all its boundary points it contains all its limit points. Conversely, let E contain all its limit points and let y be a boundary point of E. If y is a limit point of E it is in E by hypothesis. If y is not a limit point of E, some neighborhood of y contains no point of E except y itself; but then $y \in E$. Therefore E contains all its boundary points.

5.20. Let x be a limit point of E and let N_1 be a neighborhood of x. By hypothesis, N_1 contains a point y_1 of E such that $y_1 \neq x$. A neighborhood N_2 of x of radius less than $d(x, y_1)$ does not contain y_1, but does contain a point y_2 of E. And so on.

5.21. Let F be the set of limit points of E, and let x be a limit point of F. Then every neighborhood of x contains points of F, that is, limit points of E, and so contains a subneighborhood which contains points of E. Thus x is itself a limit point of E, and hence $x \in F$. That is, F contains all its limit points.

5.21a. A limit point x of the boundary B of E has in every neighborhood a point of B, hence a point of E and a point of $C(E)$; hence it is a boundary point of E, i.e., a point of B.

5.22. (a) and (c): The points of $[0, 1]$; (b): the single point 0.

5.22a. (i) The whole set.

(ii) The points $1, \frac{1}{2}, \frac{1}{3}, \ldots$ on each coordinate axis, together with $(0, 0)$.

(iii) Each radius $\theta = 1, \frac{1}{2}, \frac{1}{3}, \ldots, 0, 0 \leqslant r \leqslant 1$, consists of limit points.

5.23. Let x be a limit point of E. Every neighborhood of x contains infinitely many points of E (Exercise 5.20); since each such point is in A or in B, an infinite number must belong to one or the other of these sets.

5.24. If the set E has an empty boundary, it is both open and closed (Exercise 5.16; definition of closed).

5.25. We have to show first that a point x of the closure of E, if not in E, is a boundary point of E; we know that it is a limit point of E. Every neighborhood of x contains points of E and a point (namely, x) of $C(E)$; hence x is a boundary point of E. Now let F be the closure of E, and write $F = E \cup H$, where H consists of the limit points of E. A limit point y of F is either a limit point of E or of H. (Exercise 5.23). In the first case $y \in F$. Since H is closed (Exercise 5.21), in the second case we have $y \in H$, so $y \in F$.

5.25a. No. For example, the set $1, 2, 3, \ldots$ in R^1.

5.26. (a) $[0, 1]$; (b) $\{0, 1, \frac{1}{2}, \frac{1}{3}, \ldots \}$; (c) $[0, 1]$.

5.27. By Exercise 5.25 the closure of the neighborhood N in question is the union of N and its boundary. The boundary is the set of points y such that $d(x, y) = r$.

The statement is not true in metric spaces in general. Consider, for example, the space consisting of $[0, \frac{1}{2}] \cup [1, \infty]$ with the R_1 metric. Let N be the set of points at distance less than 1 from 0. The closure of N is the set $[0, \frac{1}{2}]$, not the set of points of the space at distance not exceeding 1 from 0.

5.28. Suppose we know that the union of n closed sets is closed. Let $F_1, \ldots, F_{n+1}$ be $n + 1$ closed sets. Then $F_1 \cup F_2 \cup \cdots \cup F_{n+1} = (F_1 \cup F_2 \cup \cdots \cup F_n) \cup F_{n+1}$, the union of two closed sets. Similarly (or by considering complements) for intersections of open sets.

5.29. The proof that the intersection of two closed sets is closed extends practically word for word.

5.30. Consider the sets in R_1, each consisting of a single rational number.

5.31. Unions of open sets are open: Consider a point x belonging to at least one of the sets of a collection of open sets. Then a neighborhood of x belongs to one of the open sets (because it is open) and therefore to the union of all the open sets.

Finite intersections of open sets are open: It is enough to consider two open sets (use induction for more). Let G_1 and G_2 be open, $x \in G_1 \cap G_2$. Then a neighborhood of x belongs to G_1, and a neighborhood of x belongs to G_2. The intersection of these neighborhoods contains a smaller neighborhood that belongs both to G_1 and G_2, and hence to their intersection.

Infinite intersections of open sets need not be open: Consider the intervals $(-1/n, 1/n)$ in R_1. Their intersection is not an open set, since it contains the single point 0.

5.32. If N_2 has any boundary points, they must belong to the set of points y such that $d(x, y) \leq r/2$; hence $N_2 \subset N_1$. If the space consists of the integers with the R_1 metric, and $r = 1$, we have $N_1 = N_2$ when $x = 0$.

6.1. A neighborhood in Ω is a single point if its radius is less than 1. Hence the sets in Ω that contain just one point are not nowhere dense.

6.2. If a set fails to be nowhere dense its closure fills some neighborhood. If the set is also closed, it coincides with its closure; hence it contains a neighborhood.

6.3. R_1, considered as a subset of R_2, is closed, and every point of R_1 is a limit point of R_1, so R_1 is perfect. But R_1 contains no neighborhood in R_2.

6.3a. Yes, because there are only countably many rational points but uncountably many points in the Cantor set. The easiest way to find an explicit irrational point is to appeal to p. 40: The Cantor set consists of base-3 "decimals" containing only the digits 0 and 2. Any such expansion that is nonterminating and nonrepeating represents an irrational number.

The point $0.77245\ldots$ is greater than $\frac{2}{3}$, so not removed in the first step of the construction of the Cantor set; less than $\frac{7}{9}$, so not removed in the second step; and so on. At the fifth step we find an interval that contains the point and is removed. Therefore the point is not in the Cantor set.

Alternatively, convert $0.77245\ldots$ to base 3 and look at the digits.

6.4. The points of the Cantor set, excluding endpoints, can be written as ternary decimals that contain no 1's and do not end in all 0's or all 2's. Supposing them enumerated as $p_1, p_2, \ldots$, we form a new number t, whose ternary digits t_n are 0 or 2 according as the nth digit of p_n is 2 or 0. Since t differs from p_n in the nth digit, it cannot occur in our alleged enumeration. This construction would fail if it happened that the nth digit of p_n were always 0 (or always 2) from some n onward. This difficulty can be avoided by renumbering the alleged enumeration before starting the construction.

6.5. The points with two rational coordinates form a countable set which is everywhere dense in R_2.

6.6. There are at least as many polynomials as there are constant terms for polynomials.

6.7. There are countably many rational constants; countably many linear polynomials with (two) rational coefficients; and so on (cf. Exercise 3.3).

6.8. If p_n is a polynomial of degree n, by approximating each coefficient within $\epsilon/(n + 1)$ by a rational number we approximate the polynomial on [0, 1] within ϵ. Use Exercise 6.7.

6.9. There are uncountably many sequences of 0's and 1's (halve the elements of the ternary expansions considered in Exercise 6.4).

6.10. 1.

6.11. Let f_x be the function defined by $f_x(t) = 0$ for $0 \leqslant t < x$, $f_x(t) = 1$ for $x \leqslant t \leqslant 1$. There are uncountably many such functions, since there is one for each x in [0, 1]. The distance between any two f_x's is 1. The rest of the proof is like that for the space m.

7.1. Let f be a continuous function on a closed bounded set E, and suppose that f is not bounded. Then there are a point x_1 such that $|f(x_1)| > 1$; a point x_2 such that $|f(x_2)| > |f(x_1)| + 1 > 2$; and so on; generally, $|f(x_n)| > n$. The Bolzano-Weierstrass theorem yields a limit point x of the set $\{x_1, x_2, \ldots\}$; $x \in E$ since E is closed; $f(x) = \lim_{n \to \infty} f(x_n)$ since f is continuous. But $\lim f(x_n)$ cannot exist since $|f(x_n)| > n$.

7.2. The set E lies in some rectangle with sides parallel to the coordinate axes; divide it into quarters by lines bisecting all its sides, and proceed as in R_1.

7.3. If some point x is in at least 3 intervals, say $E_1, \ldots, E_k$, select one whose right-hand endpoint is largest and one whose left-hand endpoint is smallest. Since both intervals contain x, they overlap, and cover the rest of the E_j, which we can then discard. Proceed similarly with another point x that is not in the two selected E_j, if there are any such points.

7.4. Let $E \supset F$, E compact, F closed. Let F be covered by open sets G. The complement $C(F)$ of F is open, and $\{G\}$ and $C(F)$ together cover E. Since E is compact, a finite subcollection drawn from $\{G\}$ and $C(F)$ covers E, and so covers F. Discarding $C(F)$ still leaves F covered.

7.5. As for Exercise 7.2.

7.5a. (a) The area of a triangle with vertices in S is a continuous function (on R_6) of the 6 coordinates of the vertices. Since S is closed and bounded in R_2, the set of coordinates of vertices is a compact subset of R_6. Hence the area attains a maximum.

(b) No. For example, the largest triangle with vertices on a circumference is the inscribed equilateral triangle.

7.6. $E_1 = \{\frac{1}{2}, \frac{1}{4}, \frac{1}{8}, \dots\}$ is covered by the open intervals $(\frac{1}{2} - \frac{1}{8}, \frac{1}{2} + \frac{1}{8}), (\frac{1}{4} - \frac{1}{16}, \frac{1}{4} + \frac{1}{16}), \dots$. No finite number of these can cover E_1, since for any finite number there would be a smallest, positive, left-hand endpoint. $E_2 = \{1, 2, 3, \dots\}$ can be covered by the intervals $(\frac{1}{2}, \frac{3}{2}), (\frac{3}{2}, \frac{5}{2}), \dots$. Any finite number of these cover only a bounded part of R_1, and E_2 is unbounded.

7.7. If a finite number of the intervals cover E, let y be the smallest left-hand endpoint that occurs. The point $y/2$ is not covered. Since E is not closed, the Heine-Borel theorem is not contradicted.

7.8. The set E is closed but not bounded.

7.9, 7.10. The Heine-Borel theorem gives sufficient conditions for a covering to be reducible to a finite covering, not necessary conditions.

8.1. Suppose that R_1 is complete and let E be a nonempty set that is bounded above. Let x_1 be an upper bound for E. If $x_1 - \frac{1}{2}$ is still an upper bound for E, call this x_2; if $x_2 - \frac{1}{3}$ is still an upper bound for E, call this x_3; and so on. Since E is not empty and $\Sigma 1/n$ diverges, we eventually reach an n such that $x_{n-1} - 1/n$ is the first number of this form that is not an upper bound for E. We now have an interval of length $1/n$ whose right-hand endpoint is an upper bound for E and whose left-hand endpoint is not. Take the right-hand endpoint as y_1. Bisect the interval and take y_2 to be either y_1 or the midpoint, whichever is an upper bound for E. Continue in this fashion; $\{y_n\}$ is a Cauchy sequence which converges to the least upper bound of E.

8.2. The *set* of all different s_n has a least upper bound L. Then, given $\epsilon > 0$, there is some s_n such that $L - s_n < \epsilon$. Since the s_n increase, $L - s_m < \epsilon$ for all $m \geqslant n$. Therefore the sequence $\{s_n\}$ converges to L.

8.3. For a Cauchy sequence $\{(x_n, y_n)\}$ in R_2, $\{x_n\}$ and $\{y_n\}$ are Cauchy sequences in R_1.

8.4. The rational points of R_1.

8.5. If $s_n \to L$, every s_n from some index on is in any given neighborhood of L. If the s_n are not all equal to L from some index on, every neighborhood of L contains one of them, other than L.

8.6. If $s_n \to L$ and $s_n \in F$, where F is a closed set, either $s_n = L$ from some point on and $L \in F$; or L is a limit point of the set of different s_n, and so $L \in F$ since F is closed.

8.7. E has a least upper bound L. There are points x_n of E such that $L - x_n < 1/n$ (not necessarily all different). Then $\{x_n\}$ has L as its limit. By Exercise 8.6, $L \in E$.

8.8. If E has only a finite number of different elements, one of them occurs infinitely often and its occurrences define the subsequence. Otherwise the elements of E have a limit point; apply the subsequence principle.

8.9. Let D be the distance in question. If G is not bounded, we can replace G by the intersection of G with a large neighborhood whose boundary is more than D away from all points of F. The distance, if attained, cannot be attained for a point not in this subset of G. Therefore we may suppose that F and G are both bounded. Take points x_n in F and y_n in G such that $d(x_n, y_n) \to D$. Using Exercise 8.7, take a convergent subsequence from $\{x_n\}$, and a convergent subsequence from $\{y_n\}$; call these $\{x_n\}$ and $\{y_n\}$ again. Then $x_n \to x$ and $y_n \to y$, and $x \in F$, $y \in G$, since F and G are closed. Also

$$d(x, y) \leqslant d(x, x_n) + d(x_n, y_n) + d(y_n, y)$$

and so $d(x, y) \leqslant D$. Since $d(x, y)$ cannot be less than D, we must have $d(x, y) = D$.

8.10. (a) Yes. (b) Not necessarily. For (b), consider the integers in R_1, with the R_1 metric. The neighborhood of 0 for which $d(0, x) < \frac{1}{2}$ is the single point 0, and its diameter is 0.

8.11. Let the diameters of E and its closure be δ and Δ, respectively; obviously $\Delta \geqslant \delta$. Choose x_n and y_n in the closure of E so that $d(x_n, y_n) \to \Delta$. If x_n is a limit point of E, there is a point x_n' of E within distance $1/n$ of x_n; then $d(x_n', y_n) \geqslant \Delta - 1/n$. Similarly if y_n is a limit point of E. (Otherwise take $x_n' = x_n$, $y_n' = y_n$.) We now have points x_n', y_n' of E with $d(x_n', y_n') \geqslant \Delta - 2/n$, and hence $\delta \geqslant \Delta$. Therefore $\delta = \Delta$.

9.1. Consider the set (x) containing only the single point x. It is nowhere dense if its closure, which is (x) again, contains no neighborhood, that is, if (x) is not a neighborhood. This will happen if the set of y with $d(y, x) < r$ contains points other than x for every positive r, and this is the case if x is a limit point of the space.

9.2. A nonempty perfect set, considered as a metric space in itself, is of second category. Since its points are all limit points, a single point of it is nowhere dense and therefore every countable subset of it is of first category. Hence no nonempty perfect set in R_1 can be countable.

9.3. This is not immediately obvious because $B =$ boundary of E is not necessarily a subset of E. We show (a) diam $E \geqslant$ diam B, (b) diam $B \geqslant$ diam E.

(a) Let $d =$ diam B. For given positive h, we can find x and y in B so that $d(x, y) > d - h$. By the definition of boundary we can find $x' \in E$ and $y' \in E$ such that $d(x, x') < h$ and $d(y, y') < h$. Then $d(x', y') \geqslant d - 3h$. Since h is arbitrary, diam $E \geqslant d =$ diam B.

(b) Since by Exercise 8.11 E and its closure Cl E have the same diameter, it will be enough to show diam $B \geqslant$ diam(Cl E). Moreover, by Exercise 5.25, Cl $E = E \cup B$. Since Cl E is closed and bounded, there are points x and y in Cl E such that $d(x, y) =$ diam(Cl E). If x and y are in B it is clear that diam $B \geqslant d(x, y) =$ diam Cl $B =$ diam E. Suppose x is not in B. Since x is not a boundary point of E it is an interior point of E. Consider a neighborhood of x, which must contain a point z such that $d(z, y) > d(x, y)$ (take a point on the line segment from y to x, extended through x). This is a contradiction of $d(x, y) =$ diam E.

10.1. If E is the set in question, it is nowhere dense unless its closure (which is E) contains a neighborhood (Exercise 6.2). If E contains a neighborhood, its complement is disjoint from this neighborhood and so not everywhere dense.

10.2. Suppose that a closed interval in R_1 is the union of a countably infinite number of disjoint closed nonempty sets E_n. By Baire's theorem one of the sets is dense in some subinterval, and, being closed, contains that subinterval. Take the largest subinterval of this kind. Then repeat the process with what remains of the original interval. We obtain a countable collection of closed intervals I_n, each belonging to one E_n, with their union everywhere dense. If I_n and I_m have a common endpoint, this endpoint belongs both to E_n and E_m, which is impossible since the E_n are disjoint. If we remove the interiors of all the intervals I_n, the remaining set H is perfect. By applying Baire's theorem to H as in example (ii), we see that the part of H in some interval J all belongs to the same E_n; in particular, so do all the endpoints of intervals I_n that are in J; hence these I_n all belong to the same E_n, and H is empty.

11.1. It is enough to show that $E \cap (a, b)$ is of measure 0 for each (a, b), since R_1 can be covered by countably many intervals. Cover E by intervals (a_n, b_n) of total length at most $q(b - a)$. Then cover each $E \cap (a_n, b_n)$ similarly. We now have E covered by intervals of total length at most $q(b_1 - a_1) + q(b_2 - a_2) + \cdots = q[(b_1 - a_1) + (b_2 - a_2) + \cdots] \leqslant q^2(b - a)$. Repeat the process; $q^n \to 0$.

12.1. (b) (domain the single point 0); (c) and (e) (domain all of R_1).

12.2. Let x be a rational number p/q. Then if $m > q$, $m! \, x$ is an even integer, $\cos m! \, \pi x = 1$, the inner limit is 1 and so $f(x) = 1$. On the other hand, let x be irrational. Then $m! \, x$ is never an integer, $|\cos m! \, \pi x| < 1$, the inner limit is 0, and $f(x) = 0$.

13.1. By the triangle inequality,

$$|f(x_0) - f(x)| = |d(x_0, y) - d(x, y)| \leqslant d(x_0, x).$$

13.2. It is enough to show that $|D(x) - D(y)| \leqslant d(x, y)$. To show this, let p be a point in E; then $d(x, p) \leqslant d(x, y) + d(y, p)$ (triangle inequality). Since $D(x) \leqslant d(x, p)$ whatever point p may be (in E), we infer that $D(x) \leqslant d(x, y) + d(y, p)$. By proper choice of p, $d(y, p)$ can be made arbitrarily close to $D(y)$. Hence $D(x) \leqslant d(x, y) + D(y)$. By interchanging x and y we also have (by the symmetry of d) $D(y) \leqslant d(x, y) + D(x)$. Thus

$$D(x) - D(y) \leqslant d(x, y),$$
$$D(y) - D(x) \leqslant d(x, y),$$

whence $|D(x) - D(y)| \leqslant d(x, y)$.

13.3. For a constant function, the image of every nonempty set is a single point.

13.4. The inverse image of a neighborhood of $f(x_0)$ that is small enough for its closure to exclude 0 is an open set. For x in a neighborhood of x_0, included in this open set, the values $f(x)$ are in the chosen neighborhood of $f(x_0)$, and so bounded away from 0.

13.5. The set of real numbers of absolute value less than $1 + |f(x_0)|$ is open; hence its inverse image is open. This inverse image is not empty, since it contains x_0, and therefore it contains a neighborhood of x_0.

13.6. Contradicting the definition of continuity tells us that there is an $\epsilon > 0$ such that for every $\delta_n > 0$ there is an x_n for which $|x_n - x_0| < \delta_n$ and $|f(x_n) - f(x_0)| \geqslant \epsilon$.

14.1. If $f + g$ and f are continuous, so is $(f + g) - f = g$. If f is not continuous, neither is $-f$, and $f + (-f)$ is the constant function all of whose values are 0; the constant function is continuous.

14.2. Sum: Given a positive ϵ, we can find δ_1 such that $|f(x) - f(x_0)| < \epsilon/2$ when $d(x, x_0) < \delta_1$, and δ_2 such that $|g(x) - g(x_0)| < \epsilon/2$ when $d(x, x_0) < \delta_2$. Take $\delta = \min(\delta_1, \delta_2)$; then $|f(x) + g(x) - [f(x_0) + g(x_0)]| < \epsilon$ when $d(x, x_0) < \delta$.

Product: f and g are bounded in a neighborhood of x_0; let M be a common bound for the values $|f(x)|$ and $|g(x)|$. Then

$$|f(x) g(x) - f(x_0) g(x_0)| \leqslant |[f(x) - f(x_0)] g(x)|$$
$$+ |f(x_0)[g(x) - g(x_0)]|$$
$$\leqslant M|f(x) - f(x_0)| + M|g(x) - g(x_0)|,$$

and the right-hand side is small when $d(x, x_0)$ is small.

Quotient: It is enough to show that $1/g$ is continous when g is continuous if $g(x_0) \neq 0$ (use the continuity of products and $f/g = f \cdot (1/g)$). By Exercise 13.4 we have $|g(x)| \geqslant m > 0$ in some neighborhood of x_0. Then for x in this neighborhood

$$\left| \frac{1}{g(x)} - \frac{1}{g(x_0)} \right| = \frac{|g(x) - g(x_0)|}{|g(x)| |g(x_0)|} \leqslant m^{-2} |g(x) - g(x_0)|,$$

and the right-hand side is small when $d(x, x_0)$ is small.

14.3. Let G be an open set; then $C(G)$ is closed. Hence if the images of closed sets are closed, the image of $C(G)$ is closed. Since f is univalent, the complement of the image of $C(G)$ is the image of G, and this complement is open (as the complement of a closed set). The converse is proved similarly.

14.4. If f does not take the value M, the function g defined by $g(x) = 1/[M - f(x)]$ is continuous on the domain of f. Accordingly, g is bounded; let G be an upper bound for $|g(x)|$. Then $1/[M - f(x)] \leqslant G$. This implies that $f(x) \leqslant M - (1/G)$, whence M is not the least upper bound for the values of f.

14.5. Let l and L be the smallest and largest points of the range. These are both values assumed by the function; it then assumes every value of the interval $[l, L]$, by the theorem just proved.

14.5a. If $f(x) \equiv 0$ there is nothing to prove. Otherwise $f(x_0) > 0$ for some x_0. Take $b > a$ so that $f(x) < \frac{1}{2}f(x_0)$ for $x \geqslant b$. Then max $f(x)$ on $[a, \infty) =$ max $f(x)$ on $[a, b]$.

14.5b. Let x_n be the last (i.e., largest) point where f takes its maximum on (n, ∞). (There is a last point since $f(x) \to 0$ and the set where $f(x) = f(x_n)$ is compact.) We clearly have $x_{n+1} \geqslant x_n$, $x_n \to \infty$.

14.6, 14.7. The range of f is the same as the range of the restriction of f to $[0, p]$. The latter function is continuous on a compact set.

$$14.8. \quad \int_x^{x+p} f(t)\, dt = \int_0^p f(t)\, dt - \int_0^x f(t)\, dt + \int_p^{p+x} f(t)\, dt$$

$$= \int_0^p f(t)\, dt - \int_0^x f(t)\, dt + \int_0^x f(t+p)\, dt$$

$$= \int_0^p f(t)\, dt,$$

since $f(t + p) = f(t)$. Alternatively,

$$\frac{d}{dx} \int_x^{x+p} f(t)\, dt = f(x + p) - f(x) = 0.$$

14.9. $\int_0^p [f(x + a) - 2f(x) + f(x - a)]\, dx = 0.$

14.10. Lévy's example was $f(x) = \sin^2(\pi x/a) - x \sin^2(\pi/a)$, $a \neq 1/n$. P. R. Halmos has pointed out that, similarly, $f(x) = g(x) - x$ is an example whenever g is periodic, continuous, and of period a, with $g(0) = 0$ and $g(1) = 1$. This figure shows an example where $\frac{1}{3} < a < \frac{1}{2}$.

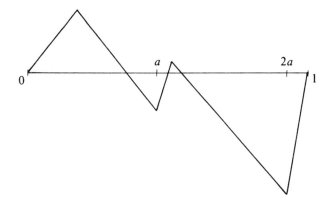

14.11. If $a = \frac{1}{2}$, our hypothesis says that f has a horizontal chord of length $\frac{1}{2}$. If $a = \frac{1}{3}$, f has either a horizontal chord of length $\frac{1}{3}$ or length $\frac{1}{2} \cdot \frac{2}{3} = \frac{1}{3}$. And so on.

14.12. Consider the function g such that $g(x) = f(x) - x$. We have $g(0) = f(0) \geq 0$, $g(1) = f(1) - 1 \leq 0$, and g continuous. Hence $g(x) = 0$ for some x.

14.12a. (a) Yes; (b) not necessarily. Let $T(t)$ be the indicated time on the erratic clock and $f(t) = t - T(t)$. Then $f(0) = 0$ and $f(24) = 0$, so f has a horizontal chord of length $1 = 24/24$, but not necessarily of any length $24/\lambda$ where $\lambda \neq$ integer; 576 minutes is $2/5$ of 24 hours. We must, however, check that our counterexample is in fact an increasing function. If $\lambda \neq 24/n$, we can take $T(t) = t + \epsilon(t\lambda \sin^2(24\pi/\lambda) - \sin^2(t\pi/\lambda))$, where ϵ is small and positive. Then $T(t + \lambda) - T(t) \neq \lambda$ and $T'(t) > 0$ if ϵ is small enough.

14.13. This is a limiting case of the theorem on simultaneous bisection of two areas (assuming that it has been proved for areas that are not convex). A direct proof is as follows. Take a point P on the curve and find another point Q so that the two arcs PQ are equal. Let $f(P)$ be the part of the area inside the curve that lies to the right of PQ. Then f is a continuous function with domain consisting of the points of the curve, since a small change in P produces a small change in Q. If P starts at P_0 and follows the curve, right and left have been interchanged when P reaches Q_0, so $f(P)$, if not originally half the area, is now on the other side of half the area and so must have been equal to half the area for some position of P.

15.1. The numbers L_n are the elements of a nonincreasing sequence. Since they are bounded they have a limit L. If n is so large that $L_n < L + \epsilon$, we have $s_k < L + \epsilon$ for $k \geqslant n$. If n is so large that $L_n > L - \epsilon$, we have some $s_k > L - \epsilon$ with $k > n$. Taking larger and larger n's we find an infinite number of s_k with the second property. Hence L has the properties defining $\limsup s_n$.

15.2. $l = \liminf s_n$ if, given any positive ϵ, we have $s_n \geqslant l - \epsilon$ if n is sufficiently large, and in addition an infinite number of s_n satisfy $s_n \leqslant l + \epsilon$. If $\{s_n\}$ is unbounded below, $\liminf s_n = -\infty$; if l fails to exist, we write $\liminf s_n = +\infty$. In the examples, l is (i) -1; (ii) $+\infty$; (iii) $-\infty$; (iv) 0; (v) 0.

15.3. We have $s_n \leqslant \frac{1}{2}\epsilon + \limsup s_n$ when $n > n_1$,

$$t_n \leqslant \tfrac{1}{2}\epsilon + \limsup t_n \text{ when } n > n_2;$$

hence

$$s_n + t_n \leqslant \epsilon + \limsup s_n + \limsup t_n \text{ when } n > \max(n_1, n_2).$$

Therefore $\limsup(s_n + t_n)$ cannot be any larger number than $\limsup s_n + \limsup t_n$. If $\lim t_n = T$, we have $s_n \geqslant \limsup s_n - \frac{1}{2}\epsilon$ for infinitely many n and $t_n \geqslant T - \frac{1}{2}\epsilon$ for all large n, whence $s_n + t_n \geqslant \limsup s_n + T - \epsilon$ for infinitely many n.

15.4. If $\epsilon > 0$ we have $s_n \leqslant L + \epsilon$ when $n > n_1$ and $s_n \geqslant L - \epsilon$ when $n > n_2$; hence $|s_n - L| < \epsilon$ when $n > \max(n_1, n_2)$.

15.5. Take $\epsilon > 0$. Then $\limsup s_n \leqslant L$ implies $s_n \leqslant L + \epsilon$ for all $n > N_1$; $\liminf s_n \geqslant L$ implies $s_n \geqslant L - \epsilon$ for $n > N_2$. Taking $N = \max(N_1, N_2)$, we have the definition of $\lim s_n = L$.

16.1. (a) $s_n \to 0$ in C. For, let ϵ be a positive number; we have $|s_n(x)| < \epsilon$ if $1 - \epsilon < x \leqslant 1$, independently of n, since $|x^n| \leqslant 1$. If $0 \leqslant x \leqslant 1 - \epsilon$, we have $|s_n(x)| \leqslant (1 - \epsilon)^n$. Therefore $\max_{0 \leqslant x \leqslant 1} |s_n(x)|$ does not exceed the larger of ϵ and $(1 - \epsilon)^n$, and the second of these numbers is the smaller if n is large enough. "Divide and rule."

(b) $\{s_n\}$ does not converge in C. For, $s_n(x) \to 0$ for each x, but $s_n(1 - 1/n) = (1 - 1/n)^n \to 1/e$, and $\max |s_n(x)|$ is no smaller than $s_n(1 - 1/n)$.

16.2. $\sup_{x \in E} |s_n(x)| \leqslant M$ by bounded convergence and therefore $\lim_{n \to \infty} s_n(x) \leqslant M$ for each $x \in E$.

17.1. The series is unchanged by termwise differentiation, so its sum $s(x)$ satisfies $s'(x) = s(x)$. Hence $s(x) = ce^x$, with $c = f(0) + f'(0) + \cdots$.

17.1a. By the M-test (p. 102), with E the set of integers, the sequence whose kth term is $\sum_{n=1}^{p(k)} f_n(k)$ converges uniformly. Then

$$\left| \sum_{n=1}^{p(k)} f_n(k) - \sum_{n=1}^{p(k)} L_n \right| \leqslant \left| \sum_{n=1}^{N} [f_n(k) - L_n] \right| + \left| \sum_{n=N+1}^{p(k)} f_n(k) \right| + \left| \sum_{n=N+1}^{p(k)} L_n \right|.$$

The last two sums can be made less than ϵ by taking N large (by uniform convergence); with N fixed, let $k \to \infty$ in the finite sum.

17.1b. For each x,

$$(1 + x/k)^k = \sum_{n=0}^{k} \binom{k}{n} \left(\frac{x}{k} \right)^n$$

$$= \sum_{n=0}^{k} \left(1 - \frac{1}{k} \right) \left(1 - \frac{2}{k} \right) \cdots \left(1 - \frac{n-1}{k} \right) \frac{x^n}{n!} ;$$

$$M_n = |x|^2 / n!.$$

17.2. For example, $f_n(x) = 0$ except for $0 < x < 2/n$, $f(1/n) = n$, $f(x)$ linear in $(0, 1/n)$ and $(1/n, 2/n)$.

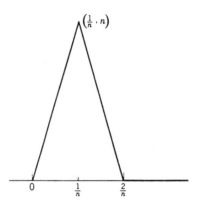

18.1. For example, $f(x, y) = \sin 2\theta$ where $x = r \cos \theta$, $y = r \sin \theta$.

19.1. No. If f is unbounded, there is a sequence $x_n \to 0$ such that $f(x_n) > n$ (or $< -n$). If f is uniformly continuous, then given $\epsilon > 0$ there is a $\delta > 0$ such that $|f(x) - f(y)| < \epsilon$ whenever $|x - y| < \delta$. In the interval from 0 to min (ϵ, δ) consider $x =$ the first x_n for which $f(x_n)$ has some value V, and $y =$ the first x_n for which $f(x_n) > V + 2\epsilon$. Then $|f(x) - f(y)| > 2\epsilon$, contradicting uniform continuity.

19.2. Yes. Since $f(x) \to L$ as $x \to +\infty$, we have $|f(x) - L| < \epsilon/2$ for $x > N$. For x and $y \leq 2N$, $|f(x) - f(y)| \leq \epsilon$ when $|x - y| < \delta$. For x and $y > N$, $|f(x) - f(y)| \leq |f(x) - L| + |f(y) - L| < \frac{1}{2}\epsilon + \frac{1}{2}\epsilon = \epsilon$. Hence $|f(x) - f(y)| \leq \epsilon$ for all positive x and y.

19.3. Given $\epsilon > 0$ we have $|f(x) - x - L| < \epsilon$ if $x > N$. Also if $h > 0$, $|f(x + h) - (x + h) - L| < \epsilon$ if $x > N$;

$$|f(x + h) - f(x)| = |f(x + h) - (x + h) - L|$$

$$+ |f(x) - x - L| + h$$

$$\leq 2\epsilon + h < 3\epsilon \quad \text{if } h < \epsilon.$$

This shows uniform continuity for $x > N$. On the compact interval $[0, N]$, f is uniformly continuous since f is continuous.

19.4. No. $f(x) = \sin x$ is a counterexample.

20.1. Since

$$\int_R^{x+y+R} = \int_R^{R+y} + \int_{R+y}^{x+y+R},$$

we get $\phi(x + y) = \phi(y) + \phi(x)$. But ϕ is a limit of continuous functions, so it has the form $\phi(x) = ax$.

21.1. If $f'_+(x)$ exists (finite), $\lim_{h\to 0+} [f(x + h) - f(x)]/h$ is finite, so $\lim_{h\to 0+} [f(x + h) - f(x)] = 0$. For the statement about f', drop all superscript and subscript $+$ signs in the preceding sentence.

21.2. If $f(x) = 0$ for $x < 0$ and $f(x) = 1$ for $x > 0$ and $f(0) = \frac{1}{2}$, we have $f'(0) = +\infty$.

21.3. $\epsilon(x) = [f(x) - f(a)]/(x - a) - f'(a) \to 0$.

21.4. $g(x + h) - g(x)$ might be zero for some values of h arbitrarily close to 0. From Exercise 21.3 we have

$$\phi(x + h) - \phi(x) = f(g(x + h)) - f(g(x))$$
$$= [g(x + h) - g(x)][f'(g(x)) + \epsilon(g(x))].$$

Since $g'(b)$ exists, g is continous at b; so $g(x) \leqslant g(b)$ as $x \to a$, and $\epsilon(g(x)) \to 0$. Divide by h and let $h \to 0$.

21.5. If $f_+(x) > 0$, $\liminf_{h\to 0+} [f(x + h) - f(x)]/h = \delta > 0$, so for all sufficiently small positive h we have $f(x + h) - f(x) \geqslant \frac{1}{2} h\delta$.

21.5a. (a) If $f'_+(x_0) = c$, finite, then if s is a given (small) positive number we have

$$c + s > \frac{f(x) - f(x_0)}{x - x_0} > c - s, \qquad x_0 + h > x > x_0,$$

provided h is small enough (depending on s). Then

$$(c + s)(x - x_0) + K(x - x_0) > f(x) + Kx - [f(x_0) + Kx_0]$$
$$> (c - s)(x - x_0) + K(x - x_0),$$
$$x_0 < x < x_0 + h,$$

and so $f(x) + Kx$ increases on the right at x_0 if $K > -c + s$, decreases on the right if $K < -c + s$; and s is arbitrarily small.

(b) If c is, for example, $+\infty$, then for a given (large) N, $x - x_0$ can be taken so small that

$$\frac{f(x) - f(x_0)}{x - x_0} > N, \qquad x_0 < x < x_0 + h,$$

where h may depend on N. This means that

$$f(x) + Kx - [f(x_0) + Kx_0] > (N + K)(x - x_0),$$

and $f(x) + Kx$ increases on the right at x_0 if $K > -N$.

(c) Suppose $f_K(x) = f(x) + Kx$ is monotonic on the right at x_0 for all K with at most one exception, and suppose that $f_K(x)$ increases for some K_1 and decreases for some K_2. Since it will also increase for every $K > K_1$ and decrease for every $K < K_2$, there must exist K_0 such that $f_K(x)$ increases for $K > K_0$ and decreases for $K < K_0$. If $K > K_0$, we have

$$f(x) - f(x_0) \geqslant -K(x - x_0),$$

whence $f_+(x_0) \leqslant -K_0$. Reasoning similarly for $K < K_0$, we get $f^+(x_0) \leqslant -K_0$. Hence $f'_+(x_0) = -K_0$ (finite).

(d) Suppose that $f_K(x)$ increases on the right at x_0 for all K. Then

$$\frac{f(x) - f(x_0)}{x - x_0} \geqslant -K,$$

where h depends on K. Since we can let $K \to -\infty$, letting (if necessary) $x \to x_0 +$, we get $f_+(x_0) = +\infty$, so $f'_+(x_0) = +\infty$. Similarly, if $f(x) + Kx$ always decreases on the right, then $f'_+(x_0) = -\infty$.

21.5b. If f_K is sometimes increasing and sometimes decreasing, then as in part (c) of Exercise 21.5a, it increases for $K > K_0$ and decreases for $K < K_0$. If $b > y > x > a$, we have

$$\frac{f(y) - f(x)}{y - x} > -K, \qquad K > K_0,$$

$$\frac{f(y) - f(x)}{y - x} < -K, \qquad K < K_0.$$

and hence

$$\frac{f(y) - f(x)}{y - x} = K_0.$$

Since K_0 is independent of x and y, we can fix y and obtain $f(x) = K_0 x + (f(y) - K_0 y)$. If, however, f_K is (say) always increasing, then for $b > y > x > a$

$$\frac{f(y) - f(x)}{y - x} > - K$$

for all K (independent of x and y), which is impossible (let $K \to - \infty$).

21.6. For $h > 0$, $f(x + h) - f(x) \leq 0$, whence $f^+(x) \leq 0$; for $h < 0$, $f(x + h) - f(x) \leq 0$ and $[f(x + h) - f(x)]/h \geq 0$, whence $f_-(x) \geq 0$.

21.7. Suppose $f'(a) < y < f'(b)$; apply the statement in the exercise to the function g defined by $g(x) = f(x) - yx$.

21.8. Let $g(x) = f(x + a) - f(x) - af'(x)$. Let c be a point where f attains its maximum. Then $f'(c) = 0$; therefore $g(c) = f(c + a) - f(c)$. Since $f(c + a)$ cannot exceed $f(c)$ (the largest value of f), we must have $g(c) \leq 0$. Let d be a point where f attains its minimum; in the same way, we get $g(d) \geq 0$. Since g is a derivative (because the continuous function f is the derivative of its own integral), g has the intermediate value property; therefore $g(x) = 0$ for some x.

21.9. Let $0 < x < 1$; then $g(x) = \int_{f(x)}^{f(1)} h(t) \, dt$ is unbounded as $x \to 0 +$, since $f(0 +) = 0$. The mean-value theorem gives $g(1) - g(x) = (1 - x)g'(c) = -(1 - x)h(f(c))f'(c)$, $0 < x < c$, and the left-hand side is unbounded as $x \to 0 +$. This proof requires fewer hypotheses than the following "change of variable" proof:

$$\int_0^1 h(f(x))f'(x) \, dx = \int_0^{f(1)} h(t) \, dt = + \infty;$$

the integrand has a fixed sign and is therefore unbounded.[59]

21.10. $f(x) = 1$ for $x < 0$, $f(x) = 0$ for $x \geq 0$.

21.11. We are supposing (tacitly) that $f'(x)$ exists for all x in a neighborhood of y ($x \neq y$). Then by the mean-value theorem $[f(x) - f(y)]/(x - y) = f'(t)$, with t between x and y; the right-hand side is as close as we like to $\lim_{x \to y} f'(x)$ if x is sufficiently close to y, hence so is the left-hand side. The last statement says that $f'(y)$ exists, and equals $\lim_{x \to y} f'(x)$.

22.1. Let $[a, b]$ be the interval: then $f(a) \leqslant f(x) \leqslant f(b)$ if $a \leqslant x \leqslant b$.

22.2. Let y be an interior point of the domain and let $\{x_n\}$ be an increasing sequence with limit y. Then $\{f(x_n)\}$ is an increasing bounded sequence which (Exercise 8.2) has a limit L. If $x_n < x < y$, we can find x_m so that $x < x_m < y$, and then $f(x_n) \leqslant f(x) \leqslant f(x_m) \leqslant L$. Since $f(x_n) \to L$, it follows that $f(x) \to L$.

22.3. If $f_n(x) \to f(x)$ for each x, and $f_n(x) \leqslant f_n(y)$ when $x \leqslant y$, then $f(x) \leqslant f(y)$ when $x \leqslant y$. The statement does not exclude the possibility that some f_n are increasing and others decreasing. If this happens, a finite number of exceptional f_n cause no difficulty. If there are an infinite number of both kinds, f must be both nonincreasing and nondecreasing, and so constant.

22.4. Let f have jumps of amount $1/\{n(n + 1)\}$ at the points $1/n$, and $f(0) = 0$. If $(m + 1)^{-1} < h < m^{-1}$, then

$$\sum_{k = m + 1}^{\infty} \frac{1}{k(k + 1)} = \frac{1}{m + 1},$$

whence $f'_+(0) = 1$.

23.1. Given $s_n \to s$, to show $(s_1 + \cdots + s_n)/n \to s$. By considering $s_n - s$ instead of s_n, we may assume $s = 0$. Then write

$$\frac{s_1 + \cdots + s_n}{n} = \frac{s_1 + \cdots + s_k}{n} + \frac{s_{k+1} + \cdots + s_n}{n}$$

and choose k so large that $|s_j| < \epsilon$ for $j > k$. Then

$$\frac{s_{k+1} + \cdots + s_n}{n} \leqslant \epsilon.$$

Now k is fixed, so $(s_1 + \cdots + s_k)/n < \epsilon$ if n is large enough, and $(s_1 + \cdots + s_n)/n < 2\epsilon$.

INDEX